INSTANT REFERENCE
WAR & WARFARE

TEACH YOURSELF®

For UK orders: please contact Bookpoint Ltd, 78 Milton Park, Abingdon, Oxon OX14 4TD.
Telephone: (44) 01235 400414, Fax: (44) 01235 400454. Lines are open 9.00–6.00, Monday
to Saturday, with a 24-hour message answering service.
E-mail address: orders@bookpoint.co.uk

For USA and Canada orders: please contact NTC/Contemporary Publishing, 4255 West
Touhy Avenue, Lincolnwood, Illinois 60646-1975, USA. Telephone: (847) 679 5500,
Fax: (847) 679 2494.

Long renowned as the authoritative source for self-guided learning – with more than 40
million copies sold worldwide – the *Teach Yourself* series includes over 200 titles in the fields
of languages, crafts, hobbies, business, computing and education.

British Library Cataloguing in Publication Data
A catalogue record for this title is available from the British Library.

Library of Congress Catalog Card Number: On file

First published in UK 2000 by Hodder Headline Plc, 338 Euston Road, London NW1 3BH.

First published in US by NTC/Contemporary Publishing, 4255 West Touhy Avenue,
Lincolnwood (Chicago), Illinois 60646-1975, USA.

The 'Teach Yourself' name and logo are registered trademarks of Hodder & Stoughton.

Copyright © 2000 Helicon Publishing Ltd

In UK: All rights reserved. No part of this publication may be reproduced or transmitted in
any form or by any means, electronic or mechanical, including photocopying, recording, or
any information storage or retrieval system, without permission in writing from the publisher
or under licence from the Copyright Licensing Agency Limited. Further details of such
licences (for reprographic reproduction) may be obtained from the Copyright Licensing
Agency Ltd, 90 Tottenham Court Road, London W1P 9HE.

In US: All rights reserved. No part of this publication may be reproduced, stored in a retrieval
system, or transmitted in any form or by any means, electronic, mechanical, photocopying, or
otherwise, without prior permission of NTC/Contemporary publishing.

Picture credits: Hulton Getty Picture Library 5, 19, 60, 120, 142, 164, 174, 195; Lionheart
Books 7, 39, 81, 85, 87, 115, 125, 150, 189, 198, 201; Fotomas Index 17, 26, 55, 62, 69, 76,
94, 99, 107, 112, 136, 146, 159, 176.

Text editors: Michael March and Bender Richardson White
Typeset by TechType, Abingdon, Oxon
Printed in Great Britain for Hodder & Stoughton Educational, a division of Hodder Headline
Plc, 338 Euston Road, London NW1 3BH, by Cox & Wyman Ltd, Reading, Berkshire.

Impression number	10 9 8 7 6 5 4 3 2						
Year	2006	2005	2004	2003	2002	2001	2000

Contents

A–Z entries	1
Appendix	203
Noteworthy Military and Naval Commanders	203
Navy Chronology	209
Arms Control Agreements	211
Weapons Chronology	218

Bold type in the text indicates a cross reference. A plural, or possessive, is given as the cross reference, i.e. is in bold type, even if the entry to which it refers is singular.

ABM
Abbreviation for anti-ballistic **missile**, a weapon used in **nuclear warfare**.

acoustic weapons
In **World War II** (1939–45), underwater weapons fitted with a sensor to detect the sound of a ship's propellers. The sensor would steer a torpedo towards the ship or, if fitted to a **mine**, detonate it when it was close enough to the ship's hull.

Acoustic **torpedoes** were introduced by the German navy in 1943 but were never very successful – some even homed in on the submarine that launched them. The Allies sought to counter acoustic weapons by towing a noise-emitting device behind ships that attracted the torpedo or triggered the mine harmlessly.

Acre, Siege of
Unsuccessful siege, on 17 March–21 May 1799, of a seaport town in Palestine, 130 km/80 mi northwest of Jerusalem, by the French under **Napoleon** Bonaparte. It became a target for Napoleon's expansionist campaign in the Mediterranean, following his expedition to Egypt. Earlier, Acre had been the focus of many military operations during the **Crusades** of medieval times.

In 1799 the city was defended by the Turks, aided by a small

Acre Napoleon during the failed Siege of Acre.

British naval force. A French assault was beaten off, and the approach of a Syrian relief army forced Napoleon to withdraw most of his force to deal with this threat. Seven more assaults had been made without success when Napoleon departed on 21 May.

admiral

Highest-ranking naval officer. In the UK **Royal Navy** and the US Navy, in descending order, the ranks of admiral are: admiral of the fleet (fleet admiral in the USA), admiral, vice admiral, and rear admiral. Famous admirals include **Nelson, Halsey**, and **Nimitz**.

The first Englishman to bear the title of admiral was William de Leyburn, to whom Edward I granted the title Admiral of the Sea of England, in 1297.

Adrianople, Battle of

Battle for the city of Adrianople (present-day Edirne, Turkey) on 15 April 1205 between the Fourth **Crusaders** led by the Latin emperor Baldwin of Flanders, who was aiming to extend his realm into mainland Europe, and the forces of the Bulgarian tsar Kalojan, who was assisting the Greek occupiers of the city. Baldwin was captured and his forces retreated in a three-day fighting march back to the security of Constantinople's walls. This was a typical 'side-show' into which Crusading armies were often distracted from their supposed aims of recapturing the Holy Land.

Aegospotami, Battle of

Naval battle fought in 405 BC off Aegospotami (now Gelibolu on the northern shore of the Dardanelles) between the Spartans and the Athenians. The Spartan commander Lysander surprised the Athenian fleet and only nine or ten Athenian ships escaped. His victory broke Athenian naval superiority and effectively ended the **Peloponnesian War**.

aerial reconnaissance

In warfare, aerial (air) reconnaissance is used to discover the position of enemy troops, fortifications, and armaments. It was first tried using balloons in the 19th century and during **World War I** aeroplanes were used for artillery spotting. Photo-reconnaissance aircraft played an important part in intelligence-gathering and bomb-damage evaluation during **World War II**. Today robot aircraft, high-flying jets, electronics surveillance planes, and satellites in space carry out reconnaissance.

Afrika Korps

German army in the western desert of North Africa in 1941–43 during **World War II**. Commanded by Field Marshal **Rommel** it seemed, for a time, invincible. The Germans recaptured **Tobruk** and advanced over the Egyptian border in June 1942 until they were halted at **El Alamein** in November 1942. The Allies gradually drove them back and the Afrika Korps were forced to surrender in May 1943.

Agent Orange

Selective weedkiller, notorious for its use in the 1960s during the **Vietnam War** by US forces to eliminate ground cover that could protect enemy forces. It was subsequently discovered to contain a highly poisonous dioxin. Thousands of US troops, who had handled it, along with many Vietnamese who came into contact with it, later developed cancer or produced deformed babies.

Agent Orange, named after the orange stripe on its packaging, combines equal parts of 2,4-D (2,4-dichlorophenoxyacetic acid) and 2,4,5-T (2,4,5-trichlorophenoxyacetic acid), both now banned in the USA.

ORANGE OUTRAGE

- Companies that had manufactured the chemicals faced lawsuits in the 1970s.
- All the suits were settled out of court, resulting in the largest ever payment of its kind (US $180 million) to claimants.

Agincourt, Battle of

English victory over the French in France on 25 October 1415, during the **Hundred Years' War**. It is famous for the use of archers with longbows against armoured **knights**. The invading English army of about 5,700 under Henry V, which had invaded France, marched in parallel with a French force about 25,000 strong toward Agincourt, a village in the Pas-de-Calais. The French refused to accept an English offer of withdrawal, and in the ensuing battle the French lost some 9,000 troops and 2,000 prisoners were taken. The English retired to Calais with losses of about 400.

- A rain-soaked ploughed field (*see map on P. 4*) separated the two sides.
- The English drove stakes into the ground to defend their archers.
- The combined effects of archers, mud, and stakes foiled the French cavalry.

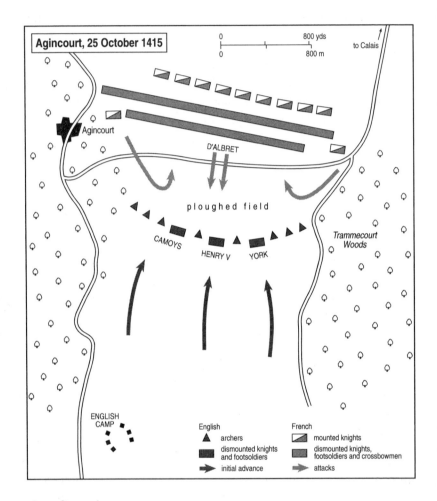

aircraft carrier

Ocean-going naval vessel with a broad, flat-topped deck for the launching and landing of military aircraft. Aircraft are catapult-launched, or take off and land on the flight-deck. The role of the carrier and its aircraft has included **aerial reconnaissance**, **torpedo**, and **bomb** operations against shipping, anti-**submarine** warfare, and air support of naval and amphibious operations. Aircraft carriers are now the equivalent of mobile airfields. The largest carriers are those of the US Navy such as *Eisenhower* (81,600 tonnes/80,294 tons, 95 aircraft) built in 1979.

Since 1945, developments have included the introduction of **jet aircraft**, the angled flight deck, mirror landing devices, and the steam catapult. Modern aircraft carriers are equipped with combinations of fixed-wing aircraft, **helicopters**, **missile** launchers, and anti-aircraft guns. The use of vertical takeoff aircraft made possible a small aircraft carrier, such as the British HMS *Invincible* (19,500 tonnes/19,200 tons).

- The first on-deck landing was on 14 November 1910 by Eugene Ely, on the USS *Birmingham*.
- The first British 'landing-on' experiments, in 1917, led to the construction of the *Argus*, the first flush-decked carrier.
- The first US aircraft carrier, USS *Langley*, was commissioned in 1922.
- The first aircraft carrier designed as such was the Japanese *Hosho* (1925).

air force
A nation's fleet of fighting aircraft and the organization that maintains them. The emergence of the aeroplane at first brought only limited recognition of its potential value as a means of waging war. Like the **balloon**, used since the American **Civil War,** it was considered a way of extending the vision of ground forces.

A unified air force was established in the UK in 1918, Italy in 1923, France in 1928, and Germany in 1935. The US air force (1947) began as the Aeronautical Division of the Army Signal Corps in 1907, and evolved into the Army's Air Service Division by 1918; by 1926 it was the Air Corps and in **World War II** the Army Air Force.

The main specialized groupings formed during **World War I** – such as combat, bombing, reconnaissance, and transport – were adapted in World War II. From 1945 to 1960, **jet aircraft** gradually superseded piston-engine aircraft. Computerized guidance systems lessened the difference between missile and aircraft, and flights of unlimited duration became possible with air-to-air refuelling. By 2000 large air forces of piloted aircraft were maintained by leading and medium-ranking military powers. **See also**: *bomb, fighter; Royal Air Force; stealth technology.*

air raid
Aerial attack, usually on a civilian target such as a factory, railway line, or communications centre, using **bombs** and **rockets**. Such attacks began just before and during **World War I** with the advent of military aviation,

but it was the development of long-range **bomber** aircraft during **World War II** that made air raids on a large scale possible.

The first long-range missiles to be used in air raids were the German **V1** 'flying bombs' and **V2** rockets directed against London during 1944.

The first air raids in World War I were carried out by **airships**, but later in the war aeroplanes were also used. Bombing was indiscriminate due to the difficulty of aiming bombs.

Many thousands died in air raids in World War II, notably during the **Blitz** on London and other British cities in 1940–41, and the firebombing of Dresden in February 1945. Air raids by both bombers and rockets have been a standard military tactic ever since.

airship

Also known as a dirigible, an airship is any aircraft that is lighter than air and power-driven. Airships were used in war for a time, but are no longer considered militarily effective.

A typical airship has an ellipse-shaped balloon that forms the streamlined envelope or hull, below it lies the propulsion system (propellers), steering mechanism, and space for crew and/or cargo. The balloon section is filled with lighter-than-air gas, such as helium or (in the past) inflammable hydrogen.

In 1852 the first successful airship was designed and flown by Henri Giffard of France. In 1900 Count (Graf) Ferdinand von **Zeppelin** of Germany designed the first rigid type, (possessing a metal framework). Airships were used by both sides during **World War I**, but were then largely replaced by aeroplanes.

In the early 1920s helium was substituted for hydrogen, reducing the danger of fire. The last and largest rigid airship was the German *Graf Zeppelin II*, completed just before **World War II** and the only zeppelin used in the war. Blimps continued to be used for coastal and antisubmarine patrol until the 1960s.

The most famous civilian airship disaster saw the destruction of the German *Hindenburg*. Forced to use flammable hydrogen by a US embargo on helium, it exploded and burned at the mooring mast at Lakehurst, New Jersey, USA, in 1937.

Alamein, El, battles of

Two battles of **World War II** that took place in the western desert of Egypt. In the first (1–27 July 1942), the British 8th Army under Auchinleck held off the German and Italian forces under **Rommel**; in the second (23 October–4 November 1942), **Montgomery** defeated Rommel, in a turning-point victory of the war.

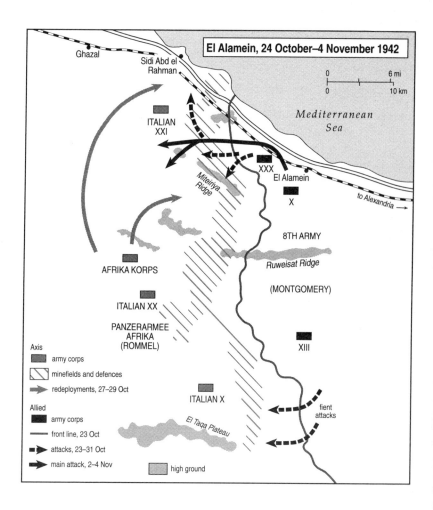

The first battle did not produce a clear winner, but was strategically vital. Rommel attacked, but Auchinleck kept him at bay. This check prevented the German and Italian armies advancing farther into Egypt.

In the second battle (*see map on p.7*), Montgomery began with a diversionary attack in the south, while the main attack in the north tried to create a gap for British armoured divisions through German minefields. An Australian attack along the coastal road on 26 October diverted **Axis** forces while a fresh attack further south developed into a major **tank** battle. By 3 November Rommel had only 30 serviceable tanks. He withdrew while the British were hampered by rain and a shortage of fuel. The defeat was a crushing blow for the Axis in North Africa.

Alamo, the
Mission fortress in San Antonio, Texas, USA. It is one of the most famous battle sites in American history, famous for the **siege** of 23 February–6 March 1836 by Gen Santa Anna and 4,000 Mexicans during the war with Texas. The Mexicans killed the garrison of about 180, including the frontiersmen Davy Crockett and Jim Bowie. The Alamo is now preserved as a US national monument.

Alexander the Great (356–323 BC)
Most famous conqueror of the ancient world, the king of Macedonia, who conquered Greece in 336 BC. He defeated the Persian king Darius in 333 at the battle of **Issus**, then moved on to Egypt where he founded Alexandria. He defeated the Persians again at **Gaugamela** (Arbela) in Syria in 331. He led his army east into Afghanistan and then into India (*see* map), where he fought one of his fiercest battles near the river Hydaspes (now Jhelum) against King Porus. His men then refused to go farther and he turned back. Alexander died in Babylon of a malarial fever.

> ❝ I will not steal a victory. ❞
>
> **Alexander the Great**, refusing to take the Persian army by surprise in 331 at Gaugamela.

Alexander, Harold Rupert Leofric George (1891–1969)
British **field marshal**, a successful commander in **World War II**. He was appointed 1st Earl Alexander of Tunis in 1952. In World War II he was the

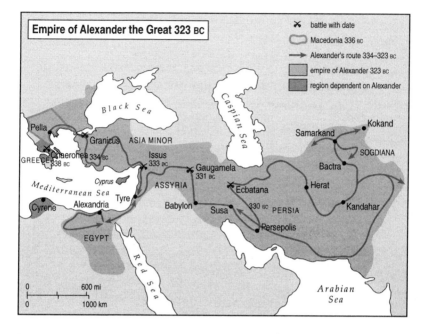

last person to leave in the evacuation of **Dunkirk** in 1940. In Burma (now Myanmar) he fought a delaying action for five months against superior Japanese forces. In 1943 he became deputy to **Eisenhower** in charge of the Allied forces in Tunisia, and later supreme Allied commander in the Mediterranean.

Alfred the Great (c. 849–c. 901)

Anglo-Saxon king in 871–899 who defended England against Danish invasion and founded the first English **navy**. His skill as a military commander first came to light at the Battle of Ashdown in 871. In 878 Alfred was forced to retire to the stronghold of Athelney, from where he emerged to win victory at Edington (Wiltshire) in 878. This battle secured the survival of the kingdom of Wessex, and his peace treaty with the Danish king Guthrum in 886 established a boundary between the Danelaw and the Saxons to the west.

- Alfred divided his levies into two parts with one half at home and the other on active service, to maintain a standing army.
- He began to build burhs (fortified towns) to form the basis of an organized defensive system.

- Alfred built a fleet of improved ships manned by Frisians and successfully challenged the Danes at sea.

> ❝ A king's raw materials and instruments of rule are a well-peopled land, and he must have men of prayer, men of war, and men of work. ❞
>
> **Alfred's** translation of Boëthius' *Consolation of Philosophy II.*

Algiers, Battle of

The city of Algiers was the focal point for a bitter war between the Algerian nationalist population and the French colonial army and French settlers in 1954–62. The war ended hopes of integrating Algeria more closely with France, and culminated in Algerian independence from French rule in 1962.

Allenby, Edmund Henry Hynman (1861–1936)

British **field marshal**, famous for his campaign in the Middle East in 1917–19 during **World War I**. Earlier he served in France, fighting at the battles of **Mons** and **Ypres** in 1914 and at **Arras** in 1917 before taking command of the British forces in the Middle East. He captured Gaza, Beersheba and, in 1917, Jerusalem. His defeat of Turkish forces at Megiddo in Palestine in September 1918 was followed by the capitulation of Turkey. He was high commissioner in Egypt in 1919–35. He was appointed Viscount in 1919.

Alma, Battle of the

Battle in the **Crimean War** in which 63,000 British, French, and Turkish troops assaulted Russian forces outside Sevastopol on 20 September 1854. A Russian force of 40,000 was defeated, but was able to hold off an attack on Sevastopol itself.

Allied troops under Lord Raglan and Marshal Jacques St Arnaud were intent upon attacking Sevastopol; the Russian commander Gen Alexander Menshikov placed his troops on high ground close to the River Alma. The British and part of the French made a frontal advance, while the remainder of the French and the Turks outflanked the Russians, who fell back into Sevastopol. About 9,000 troops died in the battle, 6,000 of them Russians.

BRITISH BLUNDER

- While French forces in the centre moved to the flank, the British were supposed to complete the frontal attack.
- Due to confused orders, part of the British line retired while the remainder advanced. The blunder was corrected in time to avoid disaster.

American Civil War *see* **Civil War, American**.

American Revolution

War of independence 1775–83, resulting in the establishment of the United States of America. The war began as a struggle between Americans and British, but widened into a general conflict involving other European powers.

The first casualties occurred in the Boston Massacre of 1770, when British troops opened fire on protesters. In 1775 fighting broke out at **Lexington and Concord**, and the Americans invaded Canada. George **Washington** was appointed commander-in-chief of the American forces. The first significant engagement of the conflict came at **Bunker Hill** on 17 June 1775, a battle won by the British at a great cost in casualties. The Declaration of Independence was issued in 1776, but Washington's troops still suffered defeats at the hands of Gen Howe.

The turning point came with the American victory at **Saratoga** in 1777. This encouraged France to enter the war on the American side, giving naval aid. At **Yorktown** in 1781 Washington and the Marquis De Lafayette, the French commander, besieged Lord Cornwallis' troops. Hopes of evacuation by sea were dashed for the British by the French

American Revolution *Washington accepts the flag of truce.*

victory over their fleet at **Chesapeake Bay** on 5 September. Cornwallis was forced to surrender on 19 October. By the Peace of Versailles, 3 September 1783, Britain recognized the independence of the USA.

- Spain declared war in June 1779, allowing New Orleans to be used as a base for privateers acting against British shipping.
- In 1780 Catherine II of Russia formed the League of Armed Neutrality to obstruct British sea power.
- US-built commerce raiders, such as that captained by the American folk hero John Paul Jones, provided naval support.

Anderson shelter

A simple **air raid** shelter used in the UK during **World War II** that could be erected in a garden to provide protection for a family. Tens of thousands were produced and they undoubtedly saved thousands of lives during the air raids on the UK. It was named after Sir John Anderson, UK Home Secretary 1939–40.

See also: *civil defence.*

Antietam, Battle of

Bloody but indecisive battle of the American civil war (see **Civil War, America**) on 17 September 1862 at Antietam Creek, off the Potomac River. Maj-Gen George McClellan of the Union blocked the advance of the Confederates under Gen Robert E **Lee** on Maryland and Washington, DC. This battle paved the way for Abraham Lincoln's proclamation of emancipation, and also persuaded the British not to recognize the Confederacy.

- Antietam was the bloodiest battle of the Civil War.
- The Confederates lost 2,700 dead, 9,000 wounded, and 2,000 missing out of 51,844 troops.
- The Union forces lost 2,100 dead, 9,500 wounded, and 750 missing out of 75,316 troops.

anti-submarine warfare

Methods used to attack and defend against **submarines**. They include: **missiles**, **torpedoes**, depth charges (explosive devices set to explode at a fixed depth in the water), bombs, and direct-fire weapons from ships, other submarines, or aircraft. **Frigates** are the ships most commonly used to engage submarines. Submarines carrying nuclear missiles are tracked and attacked with 'hunter-killer', or attack, submarines, usually nuclear-powered.

Anzio, Battle of
In **World War II**, the beachhead invasion of Italy on 22 January–23 May 1944 by Allied troops. Failure to use information gained by deciphering German codes led to Allied troops being temporarily stranded following the German attacks. They were held on the beachhead for five months before the breakthrough, which came after **Cassino** allowed the US 5th Army to dislodge the Germans from the Alban Hills, thus allowing the Anzio force to advance on Rome.

Arab–Israeli Conflict
Series of wars between Israel and various Arab states in the Middle East.
First Arab–Israeli War (15 May 1948–24 March 1949)

As soon as the independent state of Israel was proclaimed, Arab forces invaded it. The war ended in Israeli victory. Israel retained the western part of Jerusalem, Galilee, and the Negev, and went on to annex territory until it controlled 75% of what had been Palestine under British mandate.

Second Arab–Israeli War (29 October–4 November 1956)

After Egypt had taken control of the Suez Canal and blockaded the Straits of Tiran, Israel, with British and French support, invaded and captured Sinai and the Gaza Strip. Israel withdrew its forces after the entry of a United Nations (UN) force in 1957.

Third Arab–Israeli War, the Six-Day War (5–10 June 1967)

Egypt (United Arab Republic) introduced troops into Sinai. Within six days Israel's armed forces had captured the Golan Heights from Syria; the eastern half of Jerusalem and the West Bank from Jordan; and, in the south, the Gaza Strip and Sinai peninsula as far as the Suez Canal.

Fourth Arab–Israeli War, the October War or Yom Kippur War (6–24 October 1973)

Israeli forces were taken by surprise on the Day of Atonement, a Jewish holy day, when the Suez Canal was crossed by Egyptian forces, who made initial gains. The Syrians lost ground in the north.

Fifth Arab–Israeli War (6 June 1982–February 1985)

Israel invaded Lebanon. The Palestine Liberation Organization (PLO) and Syrian forces were evacuated (mainly to Syria) on 21–31 August. In February 1985, Israel withdrew, but maintained a 'security zone' in South Lebanon.

> **PEACE**
>
> - Hope of a settlement emerged in 1993 with the signing of an Israeli–PLO preliminary peace accord by Israeli prime minister Yitzhak Rabin and PLO leader Yassir Arafat.
> - Further peace negotiations followed in the 1990s between Israel, the Palestinians, Egypt, and Syria, but a state of tension remained in border areas.

Arctic convoys

In **World War II**, a series of supply **convoys** sailing from the UK to the USSR around the North Cape to Murmansk, commencing October 1941. The natural hazards of Arctic waters were added to by attacks from German surface ships, and aircraft operating from bases in northern Norway. Casualties were often heavy.

In spite of severe losses, the Arctic convoys delivered, 356,000 trucks, 50,000 jeeps, 1,500 locomotives, 9,800 freight wagons and thousands of tanks and aircraft in the course of the war.

armed forces

The military forces of a state, operating on land, sea, and in the air. They include the **air force**, **army**, **navy**, and **marines**. In the UK, all members of the armed forces are professionals, with no conscript element. In most other countries, there is an element of **conscription**.

In many armed forces, there is also a volunteer or reserve element, such as the UK Territorial Army, which acts as a general reserve for the army.

armour

Body protection worn in battle, of ancient origin. Chain mail was developed in the Middle Ages but the craft of the armourer in Europe reached its height in the 15th century.

The Greeks, Romans, and other ancient warriors wore plate armour of bronze, iron, and leather shaped to the body. **Shields** and **helmets** were also used. Later, Roman soldiers wore chain mail, a form of armour also worn by the Franks and the Vikings. Chain mail is made from interlocking rings. From the Norman Conquest of England (1066) conical helmets with nose-

ARMOURED FIGHTING VEHICLE · 15

guards were worn, together with mail shirts and leggings. A padded garment underneath gave added protection.

During the 13th century plate armour reappeared, and by the 15th century, the man-at-arms was encased in plate, articulated by rivets and straps. The use of **firearms** meant that full armour was rarely worn after 1650.

Armour reappeared in the steel helmet introduced in 1915 during **World War I**, and in the **tank**. The term armour also refers to a mechanized armoured vehicle, and to defensive shielding on tanks and ships, using steel and composite materials, such as ceramics.

BULLET PROOF

- A World War II **flak** jacket covered back and front from neck to waist. Made of fabric, it had squares of 20-gauge manganese steel sewn in.
- It could resist a .45 bullet at 9 m/30 ft.
- Contemporary bulletproof armour uses nylon and fibreglass, and is often worn beneath clothing.

---

armoured car

Wheeled, **armoured fighting vehicle**, usually armed with machine guns and sometimes a light gun. Its high speed and manoeuvrability allows it to undertake reconnaissance or support missions in situations unsuitable for **tanks**. Armoured cars were particularly effective in the North African campaigns of **World War II**.

Modern vehicles including light reconnaissance tanks, and other specialized armoured vehicles such as the armoured ambulance and armoured personnel carrier extended the traditional role of the armoured car.

armoured fighting vehicle (AFV)

A powered vehicle using wheels or chain tracks for motion, and mounting armour plate for protection against small arms and **artillery** fire, **mines**, and **grenades**. **Machine guns** or automatic cannon, **missiles**, or main-armament artillery are usually fitted to the vehicle. AFVs include **tanks**, **armoured cars**, and **armoured personnel carriers**.

The start of **World War I** saw the Allies using private touring cars as fighting vehicles. The need for a vehicle invulnerable to machine-gun fire, and which could cross lines of barbed wire and trenches was recognized by Lt Col Ernest Swinton and Winston **Churchill**. This led to the development of the tank.

In 1482 Leonardo da Vinci devised a fighting vehicle in which soldiers, housed within an armoured shell, provided their own power by turning cranks to drive four road wheels. In addition to being protected as they moved to their objective, the occupants were able to fight through ports cut in the vehicle's sides.

---

armoured personnel carrier (APC)
Wheeled or tracked military vehicle with light to medium **armour** protection used to carry infantry into and out of battle; it carries up to ten people. Many APCs are amphibious and most mount light weapons for close defence and support.

armoured train
Railway train protected by **armour** and usually carrying several guns. They were extensively used in **World War I** to protect vulnerable supply lines by patrolling or escorting supply trains, particularly by the Russian Army, which had long stretches of railway line liable to attack from roving German **cavalry** patrols.

arms trade
Sale of **weapons** from a manufacturing country to another nation. Nearly 56% of the world's arms exports end up in Third World countries. The USA is the top international weapons exporter, with global arms sales in 1996 at $31.8 billion/£19.8 billion, out of which sales to developing countries (64%) totalled $19.3 billion. Of these, the biggest customers were India, Saudi Arabia, South Korea, and Indonesia. The UK is the second biggest arms exporter, followed by Russia.

army
Organized military force of a nation, maintained by taxation, and raised either by **conscription** (compulsory military service) or voluntarily (paid professionals).

The first regular army appeared in about 1600 BC, when the Egyptian ruler,

Amosis, formed a force to fight off invaders. The Spartans trained men from childhood for service as a heavily armed infantryman, or **hoplite**. Roman armies subjected all male citizens to military service in **legions** of 6,000 men.

Most armies of ancient times were made up of **infantry** and **cavalry**. Medieval monarchs relied upon mounted men-at-arms, or **knights**, who in turn called on serfs from the land. By the 16th century the **musket**, pike, and **fortifications** combined against the knight.

The emergence of the European nation-state saw the growth of standing armies, which trained in **drills**. In the period 1792–1815, **Napoleon's** organization of his army into autonomous **corps**, and his use of mobile **artillery**, was a major step forward. In the 19th century army officers were professionally trained.

ARMIES ON THE MOVE

- In the 19th century, the **semaphore** telegraph and observation **balloons** increased the commander's ability to observe enemy movements.
- The railway revolutionized deployment of forces, supply, and evacuation of casualties.
- The American **Civil War** has been called the Railway War.

By 1914 European armies of three million were in existence, based on conscription. Breech-loading **rifles** and **machine guns** made for a higher casualty rate on the battlefields of **World War I**. Motor vehicles replaced horses. The **tank** and the radio were vital to the evolution of armoured warfare, and were developed by the Germans as **Blitzkrieg** in **World War II**.

The advent of tactical nuclear and chemical weapons brought new challenges for army commanders. From the 1960s there were developments in tanks and antitank weapons, mortar-locating **radar**, and heat-seeking **missiles**.

With the ending of the **Cold War** in the 1990s, the US and the former Soviet and European armies were substantially cut. Army tasks now include peace-keeping, as in Kosovo, with forces able to respond quickly to sudden crises.

Arnhem, Battle of

In **World War II**, airborne operation by the Allies, on 17–26 September 1944, to secure a bridgehead over the Rhine, opening the way for a thrust towards the Ruhr and a possible early end to the war. It failed in its objective.

Arnhem itself was to be taken by the British while US troops were assigned bridges to the south. Unfortunately, two divisions of the German SS **Panzer** Corps were refitting in Arnhem and penned the British troops in. Meanwhile the US force captured the bridge at Nijmegen but were unable to secure the bridge at Elst. Despite the arrival of Polish reinforcements on 21 September, the Allied commander **Montgomery** ordered a withdrawal four days later.

Arnold, Benedict (1741–1801)

General during the **American Revolution**, who won the Battle of **Saratoga** in 1777, but is remembered as a traitor to the American side. Arnold was one of the most trusted officers of **Washington**, but felt bitter at being passed over for promotion. He sent information to the British, ultimately offering to surrender the strategic post at West Point to them. Maj John André, a British soldier, was sent to discuss terms with Arnold but was caught and hanged as a spy. Arnold escaped, and was given a British army command.

Arras, Battle of

Costly if effective battle of **World War I**, in April–May 1917; a British attack on German forces in support of a French offensive, which was only partially successful on the **Siegfried Line**. British losses totalled 84,000 as compared to 75,000 German casualties.

Arras, Battle of

In **World War II**, Allied attack on German forces holding the French town of Arras on 21 May 1940 during the German invasion of France. **Rommel's** report of being attacked by 'hundreds of tanks' led to a 24-hour delay in the German advance which gave the British time to organize their retreat through **Dunkirk**. A hastily assembled force of British and French **tanks** and **infantry** launched a counterattack against Rommel's 7th **Panzer** Division. The attack caused part of the German force to panic and run. Rommel eventually beat off the British tanks using 88-mm anti-aircraft guns, the first time they were used in an anti-tank role (*see* **eighty-eight**).

arrow

Ancient and medieval **missile**, normally with a wooden shaft, a flight of feathers, and a tip which was either hardened by burning or had a metal head. It was shot by a

The arrows found on the Tudor warship *Mary Rose* were made mainly of poplar and about 75 cm/30 in long.

bow and could have considerable impact. An arrow was described in 1298 as having a barbed iron head, 7.5 cm/3 in long and 5 cm/ 2 in broad, a shaft of ash 85 cm/34 in long, and a flight of peacock feathers.

artillery

Military **firearms** too heavy to be carried, including **cannon** and wheeled guns. The earliest forms of artillery were ancient **siege** weapons, such as **catapults** and ballistas. **Gunpowder** revolutionized artillery from the 1300s. Modern artillery can also be mounted on tracks, ships, or aeroplanes and includes **rocket** launchers and specialized guns such as **howitzers**.

artillery *Two views of a 19th-century wheel-mounted heavy gun.*

Aspern, Battle of

Austrian victory over **Napoleon** on 21–22 May 1809, his first defeat during the **Napoleonic Wars**. Following the French occupation of Vienna, Austria, Archduke Charles gathered an army to face Napoleon's 40,000 troops on the Marchfield Plain, between the villages of Aspern and Essling. A confused battle ensued, Napoleon eventually retreating, fighting a rearguard action, to the village of Lobau.

assault ship

Naval vessel designed to land and support troops and vehicles under hostile conditions. It may carry specialist troops such as **marines** and also equipment such as **helicopters**.

Atlantic, Battle of the

During **World War II**, a continuous battle fought in the Atlantic Ocean by the sea and air forces of the Allies and Germany, to control supply routes to the UK. The Germans deployed **U-boats**, surface-raiders, mine-layers, and aircraft. The Allies destroyed nearly 800 U-boats, and at least 2,200 **convoys** of 75,000 merchant ships crossed the Atlantic, protected by Allied naval and air forces.

> **SIGNIFICANT EVENTS INCLUDED:**
>
> - the torpedoing (4 September 1939) of the liner *Athenia*, sailing from Glasgow to New York
> - sinking of the *Jervis Bay* (5 November 1940) by German warships
> - destruction of the German battleship *Bismarck* by the British (27 May 1941).

Atlantic Wall

Fortifications built by the Germans in **World War II** on the North Sea and Atlantic coasts of France, Belgium, the Netherlands, Denmark, and Norway. They proved largely ineffective against the Allied invasion of western Europe in 1944. The defences extended 2,750 km/1,700 mi from the North Cape to the Spanish frontier, but did not form a continuous wall. Permanent works were interspersed with barbed wire, **mines**, anti-tank ditches, steel underwater obstacles, gun positions, and other defensive devices.

atomic bomb

Also known as the A-bomb or atom bomb, a bomb deriving its explosive force from nuclear fission, as a result of a neutron chain reaction. It was developed by Allied scientists in the 1940s as the **Manhattan Project**, under the direction of US physicist J Robert Oppenheimer at Los Alamos, New Mexico. After one test explosion, two atomic bombs were dropped on the Japanese cities of **Hiroshima** (6 August 1945) and Nagasaki (9 August 1945); the bomb dropped on Hiroshima was as powerful as 12,700 tonnes/12,500 tons of TNT and that on Nagaskai was equivalent to 22,000 tonnes/21,650 tons of TNT. The USSR first detonated an atomic bomb in 1949 and the UK in 1952.
See also: *H-bomb; nuclear warfare.*

Austerlitz, Battle of

A battle during the **Napoleonic Wars**, fought on 2 December 1805. The

French forces of **Napoleon** defeated those of Alexander I of Russia and Francis II of Austria at a small town (Slavkov) in the Czech Republic (formerly in Austria), 19 km/12 mi east of Brno. The Austrians and Russians fell back before the French advance, sustaining heavy casualties as their troops were bombarded by French **artillery**. The battle was one of Napoleon's greatest victories – the Austrians signed the Treaty of Pressburg and the Russians retired to their own territory.

Austrian Succession, War of the

War of 1740–48 between Austria (supported by England and Holland) and Prussia (supported by France and Spain). The Holy Roman Emperor Charles VI died in 1740 and the succession of his daughter Maria Theresa was disputed by a number of European powers. **Frederick the Great** of Prussia seized Silesia from Austria. At **Dettingen**, in 1743, an army under the command of George II was victorious over the French. In 1745 an Austro-English army was defeated at Fontenoy but British naval superiority was already confirmed, with gains in the Americas and India. The Treaty of Aix-la-Chapelle ended the war in 1748.

auxiliary territorial service (ATS)

British Army unit of non-combatant women auxiliaries in **World War II**. Formed in 1939, the ATS provided cooks, clerks, **radar** operators, and searchlight operators, and undertook other light non-combat duties.

AWACS

Acronym for airborne warning and control system, a surveillance system that incorporates a long-range surveillance and detection **radar** mounted on a Boeing E-3 Sentry aircraft. It was used with great success in the 1991 **Gulf War**. **See also:** *aerial reconnaissance*.

Axis

The alliance of Nazi Germany and Fascist Italy before and during **World War II**. The Rome–Berlin Axis was formed in 1936, it became a full military and political alliance in May 1939. A ten-year alliance between Germany, Italy, and Japan (Rome–Berlin–Tokyo Axis) was signed in September 1940, and was subsequently joined by Hungary, Bulgaria, Romania, and the puppet states of Slovakia and Croatia. The Axis collapsed with the surrender of Italy in 1943, and of Germany and Japan in 1945.

Babi Yar
Ravine near Kiev, Ukraine, where more than 100,000 people (80,000 of whom were Jews, the remainder being Poles, Russians, and Ukrainians) were murdered by the Nazis in 1941. The site was ignored until the Soviet poet Yevgeni Yevtushenko wrote a poem called '*Babi Yar*' (1961) in protest at plans for a sports centre on the site.

Babur (Arabic 'lion') (1483–1530)
Zahir ud-Din Muhammad, the first Great Mogul of India from 1526. He was the great-grandson of the Mogul conqueror Tamerlane and, at the age of 11, succeeded his father, Omar Sheikh Mirza, as ruler of Ferghana (now Turkestan). In 1526 he defeated the emperor of Delhi at **Panipat** in the Punjab, captured Delhi and Agra, and established a dynasty that lasted until 1858.

Badajoz, Siege of
British victory over French forces in the **Peninsular War** in March–April 1812. Badajoz, a Spanish city some 400 km/250 mi southwest of Madrid, was an important fortress on the border with Portugal, which the Spanish surrendered to the French in February 1811. It was recaptured by the Duke of **Wellington** at a cost of some 5,000 casualties. British troops exacted harsh revenge for the heavy casualties they sustained, and looted the town for three days before order was restored.

Bader, Douglas Robert Steuart (1910–1982)
A British fighter pilot who lost both legs in a flying accident in 1931, but had a distinguished flying career in **World War II**. He was credited with 22 planes shot down (20 on his own) before being shot down and captured in August 1941. The film *Reach for the Sky* (1956) was based on his experiences. Bader was twice decorated and knighted in 1976 for his work with disabled people.

Bailey bridge

Prefabricated bridge developed by the British Army in **World War II**; made from a set of standardized components so that bridges of varying lengths and load-carrying ability could be assembled to order. Many Bailey bridges remained in place for several years after the war until replaced by permanent structures.

Baku, expedition to

In **World War I**, an unsuccessful British expedition to prevent the Germans controlling the Caspian Sea and deny the Russian Bolsheviks the use of oil wells in the area. At the request of anti-Bolshevik Russians, the British entered Baku in July 1918. Turkish forces besieged the town and the British were forced to evacuate in September 1918. The operation did, however, divert Turkish troops from key battlefronts in Palestine and Mesopotamia.

Balaclava, Battle of

A Russian attack on 25 October 1854, during the **Crimean War**, on British positions, near Balaclava (or Balaklava), a town in the Ukraine, 10 km/6 mi southeast of Sevastopol. It was the scene of the infamous **Charge of the Light Brigade**. The battle also gave its name to Balaclava helmets, knitted hoods worn here by soldiers to protect themselves against the bitter weather.

Balaclava, Battle of *Military uniforms worn by different nationalities in the Crimean war*

- The Russian army broke through Turkish lines and entered the valley of Balaclava.
- The Russians were driven back by the British Heavy Brigade.
- The Light Brigade then charged up the valley between Russian artillery, sustaining heavy casualties.
- The battle ended with the Russians retaining their guns and position.

Ball, Albert (1896–1917)

British **fighter** pilot and air ace in **World War I**. He was awarded the MC, DSO and Bar, and, posthumously, the Victoria Cross. At the time of his death in May 1917 he was credited with over 40 enemy aircraft shot down. Ball enlisted in the army in 1914 before transferring to the Royal Flying Corps.

Bannockburn, Battle of

Battle fought on 24 June 1314 at Bannockburn, Scotland, between Robert (I) the Bruce, King of Scotland, and Edward II of England. The defeat of the English led to the independence of Scotland.

- Edward II led over 2,000 knights and 15,000 foot soldiers, including about 5,000 archers.
- Bruce had only 500 light cavalry and some 7,000 foot soldiers.
- A badly executed night march to outflank a stream and pits dug by the Scots left Edward's knights in boggy ground.
- Bruce blocked the English knights' advance with pikemen, then charged and routed the English archers.

Barbarossa, Operation

In **World War II**, German code name for the plan to invade the USSR, launched on 22 June 1941. Initial progress was rapid and immense quantities of prisoners and equipment fell into German hands. However, due to interference from **Hitler**, the drive toward Moscow was slowed, and winter set in before the city could be taken. Large sections of the USSR, particularly the Ukraine, remained in German

> The Germans deployed massive resources for the Barbarossa campaign, organized in three Army Groups. Some 3,330 German **tanks** were deployed, with four **Luftwaffe** air fleets providing total air superiority.

hands until 1944 and fighting continued until then, notably the **sieges** of **Leningrad** and **Stalingrad**.

barrage balloon
Captive balloons, of teardrop shape and with fins to keep them headed into the wind, positioned around likely bombing targets to interfere with the probable flight paths of enemy aircraft. In **World War II** they were used to force enemy aircraft to stay high, so placing them at the optimum height for engagement by anti-aircraft guns.

Bataan Death March
In **World War II**, the brutal forced march of US and Filipino troops captured by the Japanese after the fall of Bataan in the Philippines, in April 1942. Following the surrender of Bataan, Gen **MacArthur** was evacuated, but Allied captives were force-marched 95 km/60 mi to the nearest railhead. Ill-treatment by the Japanese guards killed about 16,000 US and Filipino troops.

battalion
Basic personnel unit in the military system, usually consisting of four or five **companies**. A battalion is commanded by a lieutenant colonel. Several battalions form a **brigade**.

battleaxe
Weapon used in medieval times, by the Vikings, Normans, and others. Axes were used as both tools and weapons in prehistoric times, but the name battleaxe is used for a one- or two-handed weapon, later made of iron or steel. The poleaxe or halberd was an axe with a long handle used in medieval warfare by foot soldiers.

battleship
Class of large warships with the biggest guns and heaviest armour. The 20th century battleship was the descendant of the wooden 'man of war', such as **Nelson's** flagship HMS *Victory*, used at **Trafalgar** (1805), and now preserved at Portsmouth, England. These ships were replaced by the **ironclads** of the 19th century, which gave way to

Refurbished US battleships were used in the **Vietnam War** and the **Gulf War** to shell shore installations. There are now no battleships in active service.

the first **dreadnoughts** and thence to the great battleships of the early 20th century.

Probably the two greatest sea battles involving battleships were **Tshushima** in 1905, and **Jutland** in 1916. Battleships played a useful role in **World War II**, although **aircraft carriers** had by then begun to succeed them.

bayonet
Short sword attached to the muzzle of a firearm. The bayonet was placed inside the barrel of the muzzle-loading **muskets** of the late 17th century. The sock or ring bayonet, invented in 1700, allowed a weapon to be fired without interruption, leading to the demise of the pike. From the 1700s, bayonets evolved into a variety of types. They are still fitted to **rifles**, although in practice the bayonet is now rarely used in combat.

bazooka
A US 2.36-inch (59.9 mm) calibre rocket launcher fired from the shoulder. A lightweight tube with simple sights, it fires a fin-stabilized rocket containing a shaped charge warhead. The weapon's name came from its supposed similarity to a burlesque musical instrument played by Bob Burns, a US comedian.

Beachy Head, Battle of
English naval defeat in the Channel on 30 June 1690 by a French force sailing in support of a proposed Jacobite rebellion. The main English fleet missed the French fleet. The remainder of the English fleet met the French off Beachy Head (Sussex). Both fleets became becalmed, and the English sustained heavy casualties before retreating into the River Thames, leaving the French in control of the Channel.

The English admiral Byng was court-martialled but acquitted. 'While we had a fleet in being,' he said, 'they would not dare to make an attempt'. From this came the strategic doctrine of the 'fleet in being' whereby the existence of a powerful fleet becomes a deterrent in itself.

Beatty, David, 1st Earl Beatty (1871–1936)
British admiral in **World War I**. He commanded the cruiser squadron 1912–16 and bore the brunt of the Battle of **Jutland** in 1916. In 1916 he

became commander of the British fleet, and in 1918 received the surrender of the German fleet. He was created an earl in 1919.

> ❝ There's something wrong with our bloody ships today, Chatfield. ❞
>
> **Admiral Beatty**, complaining during the Battle of Jutland in 1916.

Belsen
Site of a Nazi concentration camp in Lower Saxony, Germany, which became notorious when entered by Allied troops towards the end of **World War II**. Established in 1943 it was not officially an extermination camp, but an outbreak of typhus in 1945 caused thousands of deaths. When Belsen was captured by British troops on 13 April 1945, several thousand bodies lay around the camp and the remaining inmates were barely alive. It was the first camp to be taken by the Allies and newsreel footage of the conditions appalled the general public.

Berlin, Battle of
Final battle of the European phase of **World War II**; on 16 April–2 May 1945; Soviet forces captured Berlin, the capital of Germany, seat of government, and site of most German military and administrative headquarters. **Hitler** committed suicide on 30 April as the Soviets closed in and Gen Karl Weidling surrendered the city on 2 May. Soviet casualties came to about 100,000 dead; German casualties are unknown but some 136,000 were taken prisoner and it is believed over 100,000 civilians died in the course of the fighting.

Berlin, Battle of (air)
In **World War II**, series of 16 heavy bombing attacks on Berlin by the **Royal Air Force** between November 1943 and March 1944. Some 9,111 **bomber** sorties were flown during the course of the campaign and immense damage was done to the city. Almost 600 bombers were lost during the battle, into which the Germans threw their entire air defence capability.

Big Bertha
Large German **howitzer** guns that were mounted on railway wagons during **World War I**. Although the name is commonly applied to many large-

calibre German guns, it specifically refers to just one, the 42-cm/16.5 in Krupp howitzer L/14, used to reduce the fortress of Liège and other strongpoints in 1914. The guns were named after the wife of the manufacturer, Gustav Krupp.

biological warfare

The use of living organisms, or of infectious material derived from them, to bring about death or disease in humans, animals, or plants. At least ten countries have this capability.

Biological warfare, together with **chemical warfare**, was originally prohibited by the 1925 Geneva Protocol. Nevertheless research continues; the Biological Weapons Convention of 1972 permits research for defence purposes but does not define how this differs from offensive weapons development. Advances in genetic engineering make the development of new biological weapons more likely.

bioterrorism

Use of biological weapons in terrorism. Diseases that could be employed as weapons include anthrax, plague, and botulism. The first use of biological weapons against civilians by a nonmilitary organization occurred in March 1995 when a Japanese cult, Aum Shinrikyo, used the nerve gas *sarin* in an attack on the Tokyo subway that left 12 dead and 5,000 hospitalized.

Bismarck

German **battleship** that had a brief career during **World War II**. The *Bismarck* displaced 50,968 tonnes/50,153 tons and had a top speed of 29 knots (54 kph). Launched in February 1939, it was a threat to Allied **convoys** in the Atlantic until sunk by the British navy in a sea battle in May 1941.

- *Bismarck* was attacked by *Hood* and *Prince of Wales* on 24 May. *Hood* was sunk and *Prince of Wales* withdrew.
- Torpedo bombers from the aircraft carrier *Ark Royal* attacked on 26 May and, on 27 May, *Bismarck* was hit by the guns of *King George V* and *Rodney*.
- A torpedo attack by *Devonshire* sank the ship. Only 107 of the 2,192 crew survived.

blackout

In wartime, the policy of keeping cities in darkness to conceal them from enemy aircraft at night. It was used during **air raids** in **World War II**, and for a time in Britain was regarded as a key element in civil defence precautions. People were reprimanded by wardens if lights shone through blackout curtains in house windows.

Blenheim, Battle of

British victory over the French in the War of the Spanish Succession on 13 August 1704. To forestall a French plan to join with Bavarian allies and march on Vienna, the British commander **Marlborough** and Prince **Eugène** of Savoy combined their forces to attack Bavaria. The Austro-British army met the opposing armies at Blindheim (Blenheim), northwest of Augsburg. A British drive through the French centre allowed the Austrians to make a flank attack and split the enemy forces. The battle marked a turning point in the war.

> Brig Rowe proceeded within thirty paces of the pales about Blenheim before the enemy gave their first fire, by which a great many brave officers and soldiers fell.
>
> Dr Hare's *Journal*, describing the British attack at **Blenheim**.

Blitz

The attempted saturation bombing of London by the German air force between September 1940 and May 1941 during **World War II**. Blitz is an abbreviation for Blitzkrieg.

Blitzkrieg (German 'lightning war')

Swift military campaign, as used by Germany in 1939–41, at the beginning of **World War II**. It was characterized by rapid movement by mechanized forces, supported by tactical air forces acting as 'flying artillery' and is best exemplified by the campaigns in Poland in 1939 and France in 1940.

Boer wars
See **South African Wars**.

Bofors gun

Light 40-mm/1.6-in anti-aircraft gun designed by the Bofors company of Sweden in 1929, used by almost all combatants in **World War II** and highly effective against low-flying ground attack aircraft. It fired at a rate of 120 rpm (rounds per minute).

bomb

Container filled with explosive or chemical material and generally used in warfare. There are also **incendiary bombs** and nuclear bombs and **missiles** (*see* **nuclear warfare**). Any object designed to cause damage by explosion can be called a bomb (such as car bombs and letter bombs).

Initially dropped from aeroplanes (from **World War I**), bombs were, in **World War II,** also launched by rockets such as the **V2**. The fusion or hydrogen bomb **(H-bomb)** was developed in the 1950s, and by the 1960s nuclear warheads 5,000 times more powerful than those of World War II were in being. In the 1970s laser-guidance systems were developed to guide conventional high-explosive bombs to hit small targets with accuracy.

bomber

Aircraft designed to drop **bombs** on the enemy. Aerial bombing started in **World War I** (1914–18) when the German air force carried out 103 raids on Britain, dropping 269 tonnes/273 tons of bombs, using aircraft such as the **Gotha**. In **World War II** (1939–45) more than ten times this tonnage was regularly dropped in successive nights on one target by aircraft such as the Lancaster (UK) and B-17 (USA). Modern jet aircraft use a combination of bombs and missiles, including 'smart' laser-guided bombs. These systems' effectiveness was demonstrated during the **Gulf War** of 1991.

Borodino, Battle of

French victory over Russian forces under Kutuzov on 7 September 1812 near the village of Borodino, 110km/70 mi northwest of Moscow, during Napoleon's invasion of Russia. Napoleon, with 137,000 troops and 585 guns, had advanced as far as Borodino, where a Russian force of about 110,000 had taken up a strong position. The French victory meant Napoleon was able to continue his advance on Moscow.

Borodino was one of the bloodiest battles of the **Napoleonic Wars**: the Russians lost 15,000 dead and 25,000 wounded; the French lost about 28,000, including 12 generals.

Bosworth, Battle of

Battle fought on 22 August 1485, during the English Wars of the Roses (*see* **Roses, Wars of the**). Richard III, the Yorkist king, was defeated and killed by Henry Tudor, who became Henry VII. The battlefield (*see* map) is near Market Bosworth, 19 km/12 mi west of Leicester, England.

Richard had 11,000–12,000 men and a strong position on Ambion Hill. Henry had 5,000–7,000 troops. A key point was the decision by Lord Stanley and his brother, with 8,000 men, to support Henry. After fierce

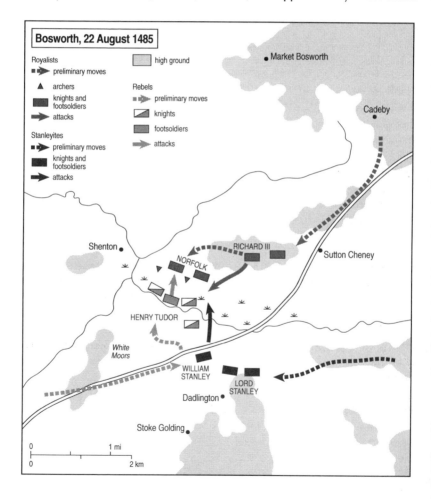

fighting, Richard charged with his cavalry and almost cut his way through to Henry before he was killed as the Stanley troops joined the fray.

bouncing bomb
Rotating **bomb** used by the British in **World War II** to attack the Ruhr dams. Designed by Dr Barnes Wallis, the bomb was slung beneath a Lancaster bomber and rotated prior to dropping from a carefully calculated height. It then rolled or 'bounced' along the surface of the water as far as the dam and sank when it came into contact with the dam wall. A depth-sensitive fuse detonated the bomb, the water in the reservoir acting as a tamping device to direct the full force of the explosion at the dam.

bow
Hand weapon used for shooting **arrows**. In its simplest form it consisted of a wooden stave with a string attached to both ends. Pulling the string bent the stave; releasing the string then shot the arrow as the stave straightened. The shortbow was a composite bow, with the stave in three pieces fixed together with glue, sinew, and horn. The **crossbow** was mechanical with a trigger. The longbow was between 1.5 m/5 ft and 1.8 m/6 ft long, usually made of yew with hemp for string. The longbow gave English armies an advantage over the French and their crossbows through much of the **Hundred Years' War**.

Boyne, Battle of the
Battle fought on 1 July 1690 in eastern Ireland, in which the exiled king James II was defeated by William III and fled to France. It was the decisive battle of the War of English Succession, ending any hopes of James's restoration to the English throne, and is enshrined in the mythology of extremist Ulster Protestantism. It took its name from the River Boyne. William sent his cavalry across the river in a frontal assault on James's army of mainly Irish and French. After fierce fighting the Irish foot soldiers broke and James's cavalry was routed. James fled to Dublin.

Bradley, Omar Nelson (1893–1981)
US general in **World War II**. In 1943 he commanded the 2nd US Corps in their victories in Tunisia and Sicily, and in 1944 led the US troops in the invasion of France. His command, as the 12th Army Group, grew to 1.3 million troops, the largest US force ever assembled.

Born in Clark, Missouri, Bradley graduated from West Point in 1915 and served in **World War I**. After World War II, he was the first chairman of the Joint Chiefs of Staff, 1949–53.

> In war there is no second prize for the runner up.
>
> **Omar Bradley**, quoted in the *Military Review*, September 1951.

Breitenfeld, Battle of

In the **Thirty Years' War**, victory of a joint Swedish–Saxon force under King **Gustavus Adolphus** over Imperial forces under Count Tilly on 17 September 1631 at Breitenfeld, about 10 km/6 mi from Leipzig. While Gustavus was negotiating alliances with Brandenburg and Saxony, Tilly sacked **Magdeburg**, which he had promised to relieve. The Saxon forces then joined the Swedes. The key moment in the battle was the Swedish capture of Imperial guns, which they fired into the rear of advancing enemy troops. Tilly's force was soon in full retreat.

Bren gun

The standard light **machine gun** of British and Commonwealth armies during **World War II**. Probably the best light machine gun used by any army during the war, it was a gas-operated weapon of .303 in calibre adopted in 1936 to replace the Lewis gun. Weighing 10 kg/22 lb, it used a 30-round magazine mounted above the gun, and fired at 500 rpm (rounds per minute).

brigade

Military formation consisting of a minimum of two **battalions**, but more usually three or more, as well as supporting arms. There are typically about 5,000 soldiers in a brigade, which is commanded by a brigadier. Two or more brigades form a **division**. A typical armoured brigade consists of two armoured battalions and one **infantry** battalion supported by an **artillery** battalion and a field-engineer battalion as well as other logistic support.

Britain, Battle of

World War II air battle between German and British air forces over Britain 10 July–31 October 1940. It was crucial in boosting British morale and in ending any remaining German plans for an invasion of Britain.

The German air force or **Luftwaffe** had about 4,500 aircraft, compared to about 3,000 for the **RAF**, which had some recently built **radar** stations for early warning. The Battle of Britain had been intended as a preliminary to the German invasion plan, which **Hitler** postponed on 17 September and abandoned on 10 October. The battle had five phases: the preliminary phase; attack on coastal targets; attack on Fighter Command bases (*see* map); and two daylight attacks on London.

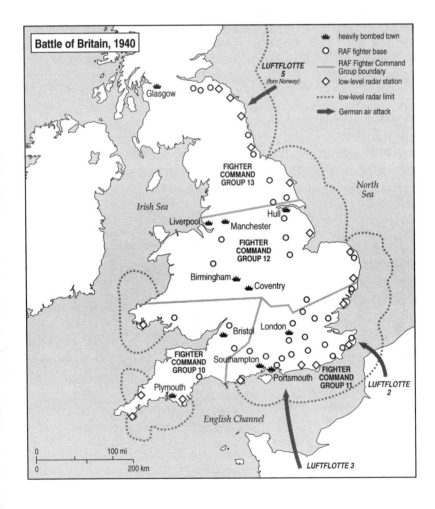

HEAVY LOSSES

- The main battle was between some 600 Hurricane and Spitfire **fighters** and the Luftwaffe's 800 Messerschmitt, 109 fighters, and 1,000 bombers (Dornier 17s, Heinkel 111s, and Junkers 88s).
- Losses in August–September were, for the RAF: 832 fighters destroyed; for the Luftwaffe: 668 fighters, and some 700 other aircraft.

British Expeditionary Force (BEF)

During **World War I** the term commonly referring to the British army serving in France and Flanders, although strictly speaking it referred only to the forces sent to France in 1914. During **World War II** it was also applied to the army in Europe, which was evacuated from **Dunkirk**, France in 1940.

Browning, John Moses (1855–1926)

US inventor of small arms, born in Ogden, Utah. The Browning pistol was much used in **World War I**. The Browning light machine gun was adopted by the US Army in 1917 and was standard issue until the early 1950s. It fired at 500 rounds per minute.

The original water-cooled Browning **machine gun** was modified to an air-cooled version after World War I and used as an aircraft and tank machine gun. A 50-in calibre version was developed as an **infantry** support and **tank** weapon.

Bulge, Battle of the

Also known as the Ardennes offensive, in **World War II**, Hitler's plan (codenamed 'Watch on the Rhine') for a breakthrough between 16 December 1944 and 28 January 1945. The Germans aimed to drive through the Ardennes, creating a 'bulge' in the Allied line. Although US troops were encircled for some weeks at Bastogne, the German counter-offensive failed. Improving weather allowed Allied air forces to play a part in the battle and by 16 January the Bulge had been eliminated. There were 77,000 Allied casualties and 130,000 German, including Hitler's last powerful **Panzer** units.

> The German offensive included a 'Trojan Horse' force of English-speaking Germans in US uniforms under Otto Skorzeny.

> All I had to do was cross the river, capture Brussels, and then go on to take the port of Antwerp. The snow was waist-deep and there wasn't room to deploy four tanks abreast, let alone six Panzer divisions. It didn't get light till eight and was dark again at four and my tanks can't fight at night. And all this at Christmas time!
>
> **Sepp Dietrich**, German SS officer, on the Battle of the Bulge

Bull Run, Battles of

In the American **Civil War**, two victories for the Confederate army under Gen Robert E **Lee** at Manassas Junction, Virginia, named after the stream where they took place, but also known as the Battle of Manassas.

In the first battle (21 July 1861) Union attacks were resisted by the brigade of Gen Thomas 'Stonewall' Jackson. Confederate reinforcements arrived and the Union troops fled in disorder. In the second battle (29–30 August 1862) Jackson again held firm until Lee and Longstreet appeared with reinforcements which, after initial delay, scattered Union forces who left behind them 14,000 dead and wounded.

> Let us determine to die here, and we will conquer. There is Jackson standing like a stone wall. Rally behind the Virginians.
>
> **Barnard Elliott Bee**, US soldier, First Battle of Bull Run.

Bunker Hill, Battle of

The first significant engagement in the **American Revolution,** on 17 June 1775, near a small hill in Charlestown (now part of Boston), Massachusetts; the battle actually took place on Breed's Hill, but is named after Bunker Hill as this was the more significant of the two. Although the colonists were defeated, they were able to retreat to Boston in good order. Gen Thomas Gage's failure to defeat them soundly resulted in his replacement as British commander.

> Men, you are all marksmen – don't one of you fire until you see the whites of their eyes.

Israel Putnam, US revolutionary soldier, giving an order at Bunker Hill.

Burma Road

Transport route running from Lashio in Burma to Kunming, China. In **World War II** it was the only route available for the Allies to send military supplies to the Chinese Army, as the Chinese coastline was inaccessible to Allied supply ships. Once the Japanese blocked the road, the only method of supply open to the Allies was to fly equipment from India to China, over the Himalayas. The road was reopened, after a year of heavy fighting by Chinese, US, and British forces, in January 1945.

Cambrai, Battles of

Two battles in **World War I** at Cambrai in northeastern France as British forces attempted to retake the town from the occupying Germans. The initial battle (20–27 November 1917) was the first in which large numbers of **tanks** were deployed, but ended with the British back where they started. From 26 August to 5 October 1918 the town was attacked as part of the push to break the **Hindenburg Line**. Specially adapted Mark V tanks crossed wide trenches and the town was recaptured on 5 October.

Cannae, Battle of

Battle fought in southern Italy in 216 BC between the Romans and Carthaginians, during the Second **Punic War**. The Carthaginian victory earned **Hannibal** immortality as a commander, while the Roman army suffered its heaviest defeat ever.

The Romans, with 80,000 men, decided to confront Hannibal in a decisive pitched battle on a narrow plain beside the river Orfanto. The Romans charged but found themselves drawn into a trap. Hannibal was able to outflank them with **infantry** and then sent in **cavalry** to attack the Romans at the rear. Between 45,000 and 70,000 Romans were killed in the battle.

cannon

Large gun developed in the later medieval period, sometimes

cannon *Early designs beginning from medieval times.*

used in battle but most often in **sieges**. Cannons were known in western Europe by the 14th century. The earliest types were made with iron rods fitted around a core and bound with rings to form a tube. By the mid-15th century wrought iron was generally used. The early weapons were loaded either by a mobile chamber with handles for lifting, or at the breech.

> The earliest known representation of a cannon in Europe is in the *Milemete Manuscript* dated 1326.

Cape Matapan, Battle of
In **World War II**, British naval victory on 28 March 1941 over an Italian force sent to disrupt Allied shipping in the Mediterranean. The Italians were intercepted just south of Crete by a British fleet under Admiral Sir Andrew Cunningham which sank the Italian cruiser *Pola*, along with the six ships sent to escort it after it had been crippled in an earlier attack.

Cape St Vincent, Battle of
During the French Revolutionary Wars, British victory over a Spanish fleet of 27 ships on 14 February 1797 off Cape St Vincent on the Portuguese coast. The British had 15 ships in tight line formation. Their triumph wrecked French plans to invade England. The two British commanders were both honoured: John Jervis became Lord St Vincent and Horatio **Nelson** gained his knighthood.

NELSON DISOBEYS

- Nelson, at the end of the British line, saw that the enemy could come at Jervis from the rear.
- In disobedience of his orders, he sailed on his own to engage them single-handedly until support arrived.
- Four Spanish ships and 3,000 prisoners were taken.

Caporetto, Battle of
In **World War I**, joint Austo-German victory over the Italian Army in October 1917. The battle took place at Caporetto, a village on the River Isonzo in northwest Slovenia. The German commander, Gen Karl von Bülow, broke through Italian lines on the Isonzo and forced an Italian retreat back onto the Piave line, where the slow Austro-German advance halted.

carrier warfare

Naval warfare involving **aircraft carriers**. Carriers were involved in several famous naval actions of **World War II**, including the sinking of the German battleship *Bismarck* in 1941. Major aircraft carrier actions took place at Taranto in 1940, the north **Atlantic** and **Pearl Harbor** in 1941, **Coral Sea**, **Midway**, Eastern Solomons, and Santa Cruz in 1942. Carriers were used extensively in the Pacific, by both the Allies and the Japanese. The US Navy deployed aircraft carriers during the **Vietnam War** and **Gulf War**.

- During World War II, 42 aircraft carriers were lost, only 3 to surface action.
- By 1945 the USA had 20 fleet, 8 light fleet, and 71 escort aircraft carriers.
- By 1945 Britain had 7 fleet, 5 light fleet, and 38 escort aircraft carriers.

Cassino, Battles of

In **World War II**, series of costly but ultimately successful Allied assaults between January and May 1944 on heavily fortified German positions blocking the Allied advance to Rome. Cassino is in southern Italy, 80 km/50 mi northwest of Naples, at the foot of Monte Cassino.

The Allies thought the Germans had fortified the monastery above the town and so it was heavily bombed on 15 February. The final battle began on 11 May 1944 when 2,000 guns bombarded the German positions. By 18 May the town and monastery were in Allied hands. Both sides sustained heavy losses in the operation.

castle

Fortified building or group of buildings, characteristic of medieval Europe. The motte and bailey castle appeared in the 11th century (the motte was a mound of earth, and the bailey a courtyard enclosed by a wall). The castle underwent many changes, largely determined by changes in **siege** tactics and the development of **artillery** and **gunpowder** weapons. Outstanding examples are the 12th-century Krak des Chevaliers, Syria (built by **Crusaders**); the 13th-century Caernarfon Castle, Wales; and the 15th-century Manzanares el Real, Spain.

See also: *fortifications*.

The main parts of a typical castle are:

- the keep, a large central tower
- the bailey or walled courtyard surrounding the keep

- embattlements through which missiles were discharged, and towers projecting from the walls
- the portcullis, a grating let down to close the main gate, and the drawbridge crossing the ditch or moat surrounding the castle.
- Barbican, a tower sometimes constructed over a gateway.

catapult

The Roman *catapulta* was an arrow- or bolt-shooting weapon, as opposed to the *ballista* (stone-throwing machine). The two words exchanged meanings during the time of the Roman Empire, probably around the 1st century AD.

Confusingly, the Greek-derived term 'catapult' also continued in use into the European Middle Ages as a general term for siege **artillery**. Roman stone-throwing machines with a single, vertical arm are usually called *onagers*.

cavalry

Mounted unit of troops deployed for their speed and manoeuvrability. Although the Egyptians appear to have used cavalry, it was not until horse-mounted troops were deployed by the Persians, that cavalry use became widespread in the ancient world. In the Middle Ages armoured **knights** were used as heavy cavalry, while **Genghis Khan's** success may be attributed to his use of light cavalry. The development of **rifles** and field **artillery** made cavalry vulnerable in the 19th century, as demonstrated by the disastrous **Charge of the Light Brigade** of 1854. After tanks and armoured cars were developed in **World War I**, most armies abandoned cavalry regiments.

Mounted troops today are retained for ceremonial roles, such as the British Blues and Royals regiment, who escort the monarch on state occasions.

Chaco War

War between Bolivia and Paraguay 1932–35 over boundaries in the north of Gran Chaco, settled by arbitration in 1938.

Chaeronea, Battle of

Battle in which the Macedonian army won a decisive victory over the confederated Greek army (mainly Athenians and Thebans) in 338 BC. The battle in central Greece marked the end of Greek independence and the start of Greek subjugation to Philip II of Macedon.

The 30,000 foot soldiers of the Macedonians held the Athenians while 2,000 cavalry, led by Philip's son **Alexander**, charged the Thebans, who broke and ran, with the exception of their famous 'Sacred Band' who fought to the last.

Chancellorsville, Battle of

In the American **Civil War**, comprehensive victory at Chancellorsville, Virginia, of Gen Robert E **Lee's** Confederate forces over Joseph Hooker's Union troops on 1 May 1863. The two armies met on the edge of a heavily wooded area known locally as 'The Wilderness'. Lee intercepted an intended Union attack on Richmond and shattered the Union army, defeating a force three times the size of his own.

Thomas ('Stonewall') Jackson, perhaps Lee's best general, was accidentally shot by one of his own men during the battle and his loss was a grave blow to the Confederate campaign.

Charge of the Light Brigade

Disastrous attack by the British Light Brigade of **cavalry** against Russian entrenched **artillery** on 25 October 1854 during the **Crimean War** at the Battle of **Balaclava**. Of the 673 soldiers who took part, there were 272 casualties.

A badly phrased order to 'prevent the enemy carrying away the guns' seems to have been intended to refer to the captured Turkish guns, but the Brigade's commander assumed his target was the Russian guns about a mile away. He led the Brigade in a charge up the length of the north valley between Russian artillery. The Brigade was saved from total destruction by French cavalry.

chariot

Ancient two-wheeled carriage, used both in peace and war by the Egyptians, Assyrians, Babylonians, Greeks, Romans, ancient Britons, and others. The war chariot was usually drawn by two horses, and sometimes had scythes attached to the wheel axles. Four horses harnessed abreast drew racing and parade chariots.

- Julius Caesar and Tacitus describe chariots being used by the British against Roman armies in the 1st centuries BC and AD.

- The most complete remains of a chariot found in Britain were at Llyn Cerrig Bach in Anglesey, Wales.

Chattanooga, Battle of

American Civil War (see **Civil War, American**) battle between 64,000 Confederate troops under Gen Braxton Bragg and 56,000 Union troops under Gen Ulysses S **Grant**, 23–25 November 1863. The Confederates broke and fell back into Georgia, putting an end to their hopes of invading Tennessee and Kentucky. The advance of Union forces through fog gave rise to the title 'The Battle of the Clouds'.

chemical warfare

Use in war of gaseous, liquid, or solid substances intended to have a toxic effect on humans, animals, or plants. Together with **biological warfare**, it was banned by the Geneva Protocol in 1925. In 1993, over 120 nations signed a treaty outlawing the manufacture, stockpiling, and use of chemical weapons. **Gas warfare** was first used as a substitute for explosives in artillery shells in **World War I** in 1915. Iraq used chemical weapons during the 1980–88 **Iran–Iraq War**.

Types of chemical weapons include:
- Irritant gases such as chlorine, phosgene (Cl_2CO), and mustard gas ($C_4H_8Cl_2S$).
- Tear gases, such as CS gas.
- Nerve gases taken into the body through the skin and lungs.
- Incapacitants that may impair vision or induce hallucinations.
- Toxins or poisons to be eaten, drunk, or injected.
- Herbicides such as the defoliant Agent Orange.
- Binary weapons that consist of two chemical components that become toxic in combination, after the shell containing them is fired.

Chesapeake, Battle of

During the **American Revolution**, French naval victory over the British off Chesapeake Bay, 5 September 1781. The French fleet had delivered reinforcements to the Marquis de Lafayette's army. A British fleet of 19 ships arrived and the French sailed out to give battle. After a mismanaged manoeuvre and some damage, the British withdrew and the French kept them at sea, allowing more reinforcements to arrive. The defeat isolated the British land force under Cornwallis who surrendered on 19 October.

Chickamauga, Battle of

Confederate victory over Union forces in the American **Civil War**, 19–20 September 1863, at Chickamauga Creek, north Georgia. Part of the Union army was routed, but Gen George Thomas rallied the troops and conducted a firm defence for six hours, before retiring in good order to Chattanooga. The Confederates besieged the town, which was eventually relieved by the Battle of **Chattanooga**.

> For beating off Confederate attacks, Gen Thomas was thereafter known as the 'Rock of Chickamauga'.

Churchill, Winston Leonard Spencer (1874–1965)

British prime minister 1940–45, an inspirational leader in **World War II**, and prime minister again 1951–55. As a soldier and military correspondent he served in India, Egypt, and South Africa before entering politics in 1900. Appointed First Lord of the Admiralty in 1911, he was forced to resign in 1915, over the **Dardanelles** disaster. He served in the trenches of France between 1915 and 16. Between 1929 and 1939, Churchill was a critic of Neville Chamberlain's appeasement policy. In May 1940 he replaced Chamberlain as prime minister, heading the wartime coalition government. The meeting of Roosevelt, Churchill, and Stalin at Yalta in 1945 planned the final defeat of Germany.

Churchill, Winston *World War II propaganda poster.*

> ❝ Never in the field of human conflict was so much owed by so many to so few. ❞
>
> **Winston Churchill**, speech of 20 August 1940, referring to the Battle of Britain.

Cid, El, Rodrigo Díaz de Vivar (c. 1043–1099)
Spanish soldier, nicknamed El Cid ('the lord') by the Moors. Born in Castile of a noble family, he fought against the king of Navarre and won his nickname *el Campeador* ('the Champion') by killing the Navarrese champion in single combat. Essentially a mercenary, fighting both with and against the Moors, he died while defending Valencia against them, and in subsequent romances became Spain's national hero.

civil defence
Also called civil protection, organized activities by the civilian population of a state to mitigate the effects of enemy attack. During **World War II** civil-defence efforts were centred on providing warning of **air raids**. Air-raid shelters were constructed and existing buildings were adapted to provide protection. Precautions against gas attacks included the free issue of gas masks. Firefighting, food, rescue, communications, and ambulance services were needed. In the 1950s the threat of **nuclear warfare** led to the building of fallout shelters in the USA, the USSR, and elsewhere.

civil war
War between rival groups within the same country. Of many examples, the most famous are the American, English, and Spanish civil wars.

Civil War, American
Also called the War Between the States, war in 1861–65 between the Southern or Confederate States of America and the Northern or Union states. The issue of slavery brought to a head the long-standing social and economic differences between North and South. The war began after Abraham Lincoln's inauguration as president in March 1861. The North had a fighting strength four times as large as that of the South, was more advanced industrially, and had the stronger navy to blockade Confederate ports. Notable commanders were Robert E **Lee** for the Confederacy and Ulysses S **Grant** for the Union.

The civil war cost over 620,000 lives. The North emerged stronger than ever, while the South was ruined. The war and its aftermath left behind much bitterness.

Main events of the American Civil War

1861 Rebel Confederate forces bombard the federal garrison at Fort Sumter, South Carolina, on 12 April, and 34 hours later the fort is surrendered. First Battle of **Bull Run**, July 21.

1862 Confederates win second Battle of **Bull Run** and Battle of **Fredericksburg**.

1863 Battle of **Chancellorsville** on 1 May is won by Confederates, but at the cost of Stonewall Jackson.

1863 **Grant** takes Vicksburg on 4 July for the Union after a siege lasting six weeks.

1863 The Confederates are decisively defeated on 1–3 July at **Gettysburg**, Pennsylvania, and **Lee** retreats into Virginia.

1863 Battle of **Chickamauga**, September, Confederate victory.

1863 Battle of **Chattanooga**, the Confederates are forced back into Georgia.

1864 Grant sets out to destroy Lee's army in Virginia. He sends **Sherman** to campaign in Georgia.

1864 The Confederate fleet is destroyed at the Battle of Mobile Bay in August.

1864 Sherman's march to the sea from Atlanta with an army of 62,000. He enters Savannah unopposed on 21 December 1864.

1864 George Thomas defeats the Confederates at the Battle of Nashville in December.

1865 Sherman begins his march back from the sea. Columbia is burned down, and Charleston was deserted by the Confederates.

1865 Union forces captures Petersburg and enter Richmond on 3 April.

1865 Lee surrenders at Appomattox Court House on 9 April.

Civil War, English

Conflict between King Charles I and the Royalists (also called Cavaliers) on one side and the Parliamentarians (also called Roundheads) under Oliver **Cromwell** on the other. Hostilities began in 1642 and a series of Royalist defeats (at Marston Moor in 1644, and then at **Naseby** in 1645) culminated in Charles's capture in 1647, and execution in 1649. The war continued until the final defeat of Royalist forces at Worcester in 1651. Cromwell then became Protector (ruler) from 1653 until his death in 1658.

Main events of the English Civil War

1642 The Royalist and Parliamentarian armies first meet at the Battle of Edgehill in October, there is no conclusive outcome.

1643	The Royalists take control of most of Yorkshire after the Battle of Adwalton Moor in June.
1644	The Parliamentarians win the Battle of Marston Moor in July.
1645	The Parliamentarian New Model Army, formed February 1645, win a resounding victory at the Battle of Naseby in June.
1648	The second phase of the Civil War begins in March with uprisings by royalist supporters in Wales and England, and a Scottish invasion.
1649	Charles I is captured, tried, and executed early in 1649. England becomes a republic.

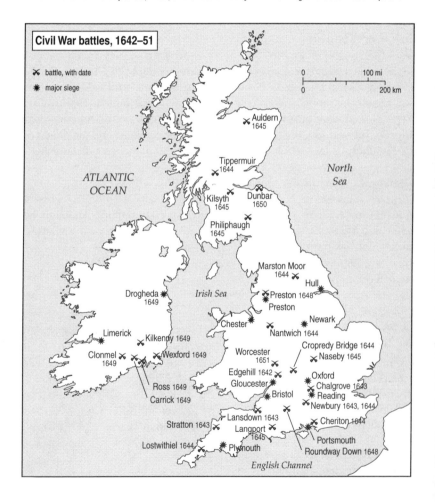

> The utterly memorable Struggle between the Cavaliers (Wrong but Wromantic) and the Roundheads (Right but Repulsive).

English writers W C **Sellar** and R J **Yeatman**, on the English Civil War, from the book *1066 and All That, A Memorable History of England* (1930)

Civil War, Spanish

War of 1936–39 precipitated by a military revolt led by Gen Franco against the Republican government. Franco's nationalists, supported by fascist Italy and Nazi Germany, seized power in the south and northwest, but were suppressed in Madrid and Barcelona by the workers' militia. The loyalists (Republicans) were aided by the USSR and volunteers of the International Brigade. The war saw the use of **air raids** against civilians, and the defeat of the Republicans by 1939 led to the establishment of Franco's dictatorship.

Clausewitz, Carl Philipp Gottlieb von (1780–1831)

Prussian officer whose book *Vom Kriege/On War* (1833) exerted a powerful influence on military strategists well into the 20th century. Although he advocated the total destruction of an enemy's forces as one of the strategic targets of warfare, his most important idea was to see war as an extension of political policy and not as an end in itself.

Clive, Robert, 1st Baron Clive (1725–1774)

British soldier and administrator who established British rule in India at a time when rivalry between the French and British East India companies was intense. He won victories over French troops at Arcot in 1751 and over the nawab of Bengal at **Plassey** in 1757. This secured Bengal for the East India Company, and made

Clive, *Robert Clive, British hero in India.*

'Clive of India' a national hero. On his return to Britain in 1766, his wealth led to allegations that he had abused his power as governor of Bengal. Although acquitted by a Parliamentary enquiry, he committed suicide.

CLIVE AND THE NAWAB

- In 1755 Siraj-ud-Daula, the nawab of Bengal, drove the British from Calcutta, and imprisoned his captives in the notorious 'Black Hole of Calcutta'.
- With a force of 1,900 men, Clive defeated the nawab's army of 34,000 outside Calcutta in February 1757.

Clontarf, Battle of
Irish victory over a Norse (Viking) invasion force on Good Friday, 23 April 1014. Although the Irish won a magnificent victory, which lifted the pagan Norse threat to Ireland, the Irish king Brian Boru and his son were both killed in the battle; Brian, being too old to fight, was slain in his tent. Details of the battle are scant, but the Norse losses were said to be 6,000 dead.

COIN
Contraction of counter insurgency, the suppression by a state's armed forces of uprisings against the state. Also called internal security (IS) or operations of counter-revolutionary warfare (CRW). An example is the operation by the Russian army in Chechnya in the late 1990s–2000.

Cold Harbor, Battle of
American **Civil War** battle near Richmond, Virginia, 1–12 June 1864, in which the Confederate army under **Lee** repulsed Union attacks under **Grant**, inflicting heavy casualties and forcing Grant to adopt a siege of Petersburg.

The Union army sustained 8,000 casualties in the first two hours on 3 June alone, and lost over 12,500 in the period of 1–12 June, against Confederate losses of fewer than 3,000. This engagement, one of the final Confederate victories of the war, kept Grant's army largely stationary until April 1865.

Cold War
Hostilities short of armed conflict, describing the relations between, on the

one side, the USA and Western Europe and on the other, the USSR and Eastern Europe, from 1945 to 1990. East–West tensions were intensified by crises such as the **Korean War** and **Vietnam War**. Both sides had large stocks of nuclear weapons, as part of plans of **deterrence**. In 1958, however, when it was agreed that it was possible to identify certain types of nuclear explosions, it became possible to impose the partial nuclear test-ban treaty signed by the USA, the USSR, and the UK in 1963. The Cold War ended with the collapse of Communist rule in Eastern Europe and the break-up of the USSR.

Colt, Samuel (1814–1862)

US gunsmith who invented the **revolver** in 1835 that bears his name. With its rotating cylinder, which turned, locked, and unlocked by cocking the hammer, the Colt revolutionized military tactics. Colt built a large factory in Hartford, Connecticut in 1854. He introduced mass-production techniques, and his weapons had interchangeable parts, making them easy to maintain and repair. During the **Crimean War** of 1853–56 he also manufactured arms in Pimlico, London. When the American **Civil War** broke out in 1861, he supplied thousands of guns to the US government.

FIRWORKS

- As a boy, Colt discovered how to fire gunpowder using an electric current.
- After a public demonstration at a mine, which covered spectators with mud, he was sent to Amherst Academy.
- As a result of a fire caused by another of his experiments, he was asked to leave!

---

combined operations

In **World War II**, operations in which all three services – **army**, **navy**, and **air forces** – were involved, notably amphibious landings and **commando** operations. The first British Chief of Combined Operations was Admiral Sir Roger Keyes in August 1940, who was killed in an abortive attempt to kidnap **Rommel**. Mountbatten succeeded him in October 1941. Combined Operations HQ was involved in planning all amphibious operations.

commando
Member of a specially trained, highly mobile military unit. The term originated in South Africa, where it referred to Boer raids against Africans and, in the **South African Wars**, against the British. Commando units often carried out operations behind enemy lines.

British Combined Operations Command units raided enemy-occupied territory in **World War II** after the evacuation of Dunkirk in 1940. Among the commando raids were those on the Lofoten Islands (3–4 March 1941), St Nazaire (28 March 1942), and **Dieppe** (19 August 1942). At the end of the war the army commandos were disbanded, but the Royal Marines carried on their role.

company
In the army, a subunit of a **battalion**. It consists of about 120 soldiers, and is commanded by a major in the British army and a captain in the US army. Four or five companies make a battalion.

In British tank and engineer battalions, a company is known as a squadron. In British artillery battalions, a company is known as a battery.

conscription
Requirement for all able-bodied male citizens (and female in some countries, such as Israel) to serve with the armed forces. It originated in France in 1792, and in the 19th and 20th centuries became the established practice in many states. In the USA it is called the draft. Modern conscription systems often permit alternative national service for conscientious objectors. In Britain conscription was abolished in 1960.

Constantinople, Siege of
Siege lasting from 1394 to 1402 in which the Ottoman sultan Bayezid I blockaded Constantinople despite Christian relief attempts. Bayezid's defeat by Timur Leng (Tamerlane) at Ankara spared the city until 1453. In 1453 it was captured by another Turkish army after nearly a year's siege and became the capital of the Ottoman Empire.

convoy system
Grouping of ships to sail together under naval escort in wartime. In **World War I** (1914–18) navy escort vessels were at first used only to accompany troopships, but the convoy system was adopted for merchant shipping when the unrestricted German **submarine** campaign began in 1917. In **World**

War II (1939–45) the convoy system was widely used by the Allies to keep the Atlantic sea lanes open.

Copenhagen, Battle of

Naval victory on 2 April 1801, during the **Napoleonic Wars**, by a British fleet under Sir Hyde Parker (1739–1807) and **Nelson**, over the Danish fleet. Nelson put his telescope to his blind eye and refused to see Parker's signal for withdrawal.

Coral Sea, Battle of

The Coral or Solomon Sea is part of the Pacific Ocean bounded by northeastern Australia. The naval battle of the Coral Sea, 7–8 May 1942, fought between the USA and Japan, mainly from **aircraft carriers**, checked the Japanese advance in the South Pacific in **World War II**. This was the first sea battle to be fought entirely by aircraft, launched from carriers, without any engagement between the actual warships themselves.

Coronel, Battle of

In **World War I**, German naval victory over a British squadron off the coast of Chile on 1 November 1914. A German squadron under Admiral Maximilian von Spee began commerce raiding in the southern Pacific. In October 1914 a small British squadron under Admiral Sir Christopher Cradock, encountered the Germans off Coronel. Cradock was outgunned and was defeated, losing two cruisers and 1,500 crew members. He was lost along with his flagship.

corps

Military formation consisting of two to five **divisions**. Its strength is between 50,000 and 120,000 people. All branches of the army are represented. A corps is commanded by a lieutenant general or, in the USA, by a three-star **general**. Two or more corps create an army group.

Coruña, Battle of

Battle of 16 January 1809, during the **Peninsular War**, to cover embarkation of British troops after their retreat to Coruña. The British commander in Portugal, John Moore, was killed after ensuring a victory over the French. British forces were arrayed in lines and the French launched a series of attacks, which were beaten off. By nightfall the French were retiring, but

Moore had been mortally wounded. The new British commander Sir John Hope withdrew by night to waiting ships and escaped.

corvette
Small and lightly armed vessel, used to escort **convoys** in **World War II**. The term, now obsolete, was revived from sailing days. Corvettes were intended as a quick and cheap counter to the threat from German **submarines**.

Crécy, Battle of
First major battle of the **Hundred Years' War**, fought on 26 August 1346. Philip VI of France was defeated by Edward III of England at the village of Crécy-en-Ponthieu, now in the Somme *département*, France. The English archers played a crucial role in Edward's victory, which allowed him to besiege and take Calais.

- Edward had some 10,000 men, including 2,000 men-at-arms, 5,000 archers and 3,000 infantry.
- Philip had perhaps 12,000 mounted men-at-arms, 6,000 Genoese crossbowmen, and 10,000 other infantry.
- The Genoese crossbowmen opened the battle, but rain had slackened their bowstrings.
- They were annihilated by Edward's Welsh bowmen who had unstrung their bows and kept the strings dry.
- In repeated charges, 1,500 French knights were killed and Philip withdrew.

Crécy, Battle of *Edward the Black Prince leading his troops.*

Crete, Battle of

In **World War II**, German operation to capture the island of Crete from the Allies during May 1941. The Germans had air superiority and were able to bomb before an air drop attack by paratroops. By 28 May evacuation of Allied troops had begun. Both sides suffered massive casualties, in particular the German airborne forces (over 50%), so much so that **Hitler** forbade any further major airborne operations.

On Crete some 3,600 Allied troops were killed and about 12,000 taken prisoner; German losses came to 6,000 killed and wounded, and some 220 aircraft lost. The Royal Navy also suffered heavy losses.

Crimean War

War of 1853–56 between Russia and the allied powers of England, France, Turkey, and Sardinia. The war arose from British and French mistrust of Russia's ambitions in the Balkans. It began with an allied Anglo-French expedition to the Crimea to attack the Russian Black Sea city of **Sevastopol**. The battles of the River **Alma**, **Balaclava** (including the **Charge of the Light Brigade**), and **Inkerman** in 1854, led to a siege which, owing to military mismanagement, lasted for a year. The scandal surrounding French and British losses through disease led to the organization of proper military nursing services by Florence Nightingale.

Crimean War, *Benjamin Disraeli (seated), leader of the Opposition, criticises the war budget of the Chancellor of the Exchequer, William Gladstone.*

Important events in the Crimean war

1853 Russia invades the Balkans (withdrawing after Austrian intervention) and sinks the Turkish fleet at the Battle of Sinope on 30 November.

1854 Britain and France declares war on Russia, invades the Crimea, and lay siege to Sevastopol (September 1854–September 1855). Battles of Balaclava, Inkerman, and the Alma.

1855 Sardinia declares war on Russia.

1856 The Treaty of Paris in February ends the war.

Cromwell, Oliver (1599–1658)
English general and politician, leader of the Parliamentary side in the English **Civil War**. He raised **cavalry** forces (later called Ironsides) that aided his victories at Edgehill in 1642 and Marston Moor in 1644, and organized the New Model Army, which he led (with Gen Fairfax), to victory at **Naseby** in 1645. He declared Britain a republic ('the Commonwealth') in 1649, following the execution of Charles I, and ruled as Lord Protector from 1653.

Cromwell, Oliver *A portrait of Oliver Cromwell in battle armour.*

❝ I had rather have a plain russet-coated captain that knows what he fights for, and loves what he knows, than that which you call a gentleman and is nothing else. ❞

Oliver Cromwell, in a letter to Sir William Spring, 1643.

crossbow
Bow with a mechanism to draw back the string and a trigger to release it, used in medieval European warfare. Bending the bow was originally done using the

hands and feet. Improvements included the use of a stirrup fitted to the stock into which the archer's foot could be placed while drawing the bow. Other devices for drawing included the pulley and the windlass. Steel bows were developed in the 15th century. The crossbow missile was known as a bolt or quarrel.

cruise missile

Long-range guided missile that has a terrain-seeking **radar** system and flies at moderate speed and low altitude. It is descended from the German **V1** of **World War II**. It has pinpoint accuracy on low-level flights after launch from a mobile ground launcher, and can be launched from the ground, from aircraft, or from a **submarine** or ship. US Tomahawk cruise missiles were used in the 1991 **Gulf War**.

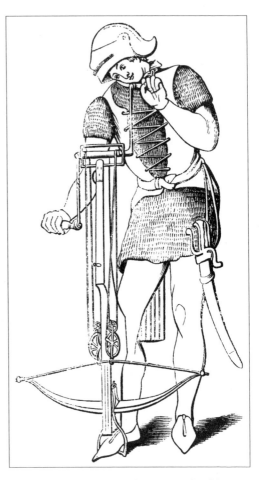

Crossbow *A medieval archer, bending his crossbow mechanically.*

Crusades

European wars against non-Christians and heretics sanctioned by the pope, in particular the series of wars of 1096–1291, undertaken by European rulers to recover Palestine from the Muslims. Motivated by religious zeal, the desire for land, and the trading ambitions of the major Italian cities, the Crusades were varied in their aims and effects.

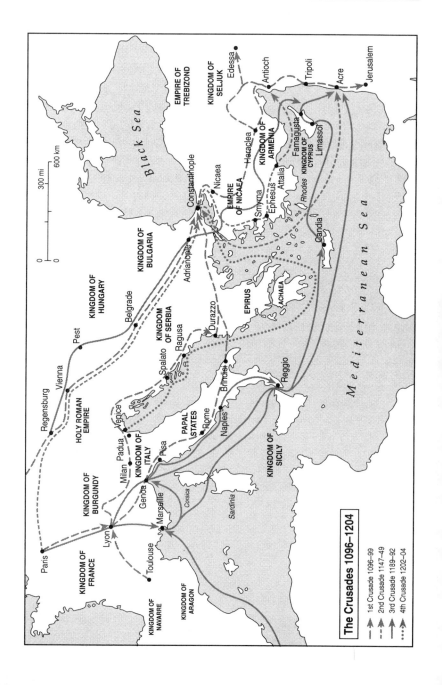

The Crusades ostensibly began to ensure the safety of pilgrims visiting the Holy Sepulchre and to establish Christian rule in Palestine and they continued for over two centuries (*see map on p.57*). Attacks were directed against Egypt and even Constantinople.

Important events in the Crusades

1096–97 The First Crusade is mounted. A Christian kingdom is established around Jerusalem. The orders of Knights Hospitaller and Knights Templar are formed to assist in the defence of Jerusalem.

1187 Muslim pressure increases with the conquests of **Saladin**, Sultan of Egypt. He defeats a Christian army at **Hattin**, and takes Jerusalem in October.

1189 Several fresh expeditions are mounted, of which the most important is the Third Crusade led by Philip (II) Augustus of France, Frederick (I) Barbarossa of Germany, and Richard (I) the Lionheart of England.

1202 The Fourth Crusade, starting from Venice, becomes involved in Venetian and Byzantine intrigue.

1204 Constantinople is sacked by the crusaders.

1249 The Seventh Crusade, led by Louis IX of France is, like an earlier expedition in 1217, directed against Egypt, and proves even more disastrous. Louis is captured with the greater part of his army, and has to pay 800,000 pieces of gold as a ransom. Even after this, he heads another crusade in 1270, but dies at Carthage.

Culloden, Battle of

Defeat in 1746 of the Jacobite rebel army of the British prince Charles Edward Stuart (the 'Young Pretender') by the Duke of Cumberland on a stretch of moorland near Inverness, Scotland. This battle effectively ended the military challenge of the Jacobite rebellion.

Although both sides were numerically equal, the English were a disciplined force, while the Jacobites forces were virtually untrained. Although the Jacobites broke through the first English line, they were caught in the **musket** fire of the second line. They retired in confusion, pursued by the English **cavalry**, which broke the Jacobite lines completely.

Dakar
Chief port of Senegal, formerly the seat of the government of French West Africa. In 1940 during **World War II** a naval action was undertaken by British and Free French forces to seize Dakar, following the French Vichy government's armistice with the **Axis** powers. British torpedo bombers attacked the French battleship *Richelieu* in July, and a full Allied naval force arrived in September. Fire was exchanged, with damage to ships of both sides before the operation was called off.

Dardanelles campaign
In **World War I**, unsuccessful Allied naval operations in 1915 against the Turkish-held Dardanelles, a narrow passage between the Mediterranean and the Sea of Marmora and thence to the Black Sea. The idea came from Winston **Churchill** in late 1914. After a series of unsuccessful attacks by warships and landing parties, a purely naval attack was abandoned in favour of a land action. The operations alerted the Turkish army who reinforced the area before the **Gallipoli** landings.

D-Day
Code-name for 6 June 1944, the day of the Allied invasion of Normandy in **World War II**. Under the command of Gen **Eisenhower**, it marked the start of Operation Overlord, the liberation of Western Europe from German occupation. The Allied invasion fleet landed on the Normandy beaches. Artificial harbours known as '**Mulberries**' were towed across the Channel so that equipment could be unloaded on to the beaches. After overcoming fierce resistance the allies broke through the German defences; Paris was liberated on 25 August.

- Five beaches – Utah, Omaha, Gold, Juno, and Sword – were the landing points.
- The landings commenced at 0630 hrs, and by midnight 57,000 US and 75,000 British and Canadian troops were ashore.

- Allied losses during the day amounted to 2,500 killed and about 8,500 wounded.
- Allied air forces flew 14,000 sorties and lost 127 aircraft.

Desert Storm, Operation

Code-name of the military action to eject the Iraqi army from Kuwait 1991. The build-up phase was code-named Operation Desert Shield and lasted from August 1990, when Kuwait was first invaded by Iraq, to January 1991 when Operation Desert Storm was unleashed, starting the **Gulf War**. Desert Storm ended with the defeat of the Iraqi army in the Kuwaiti theatre of operations late in February.

destroyer

Small, fast warship, used by many navies. Modern destroyers often carry guided **missiles** and displace about 3,700–5,650 tonnes/3,650–5,560 tons. To counter fast **torpedo**-carrying vessels, so-called 'torpedo-boat destroyers' were designed by Britain in the late 19th century. They proved so effective that torpedo-boats were more or less abandoned in the early 1900s, but the rise of the **submarine** found a new task for destroyers as **convoy** escorts and anti-submarine vessels in both **World War I** and **World War II**.

deterrence

Underlying conception of the nuclear arms race during the **Cold War**: the belief that a potential aggressor will be discouraged from launching a 'first strike' nuclear attack by the knowledge that the adversary is capable of inflicting 'unacceptable damage' in a retaliatory strike. This doctrine is widely known as that of mutual assured destruction (MAD).
See also: *nuclear warfare.*

Dettingen, Battle of

In the War of the **Austrian Succession**, battle in the Bavarian village of that name where on 27 June 1743, an army of 42,000 British, Hanoverians, and Austrians under George II defeated a French army of about 50,000. King George led his **cavalry** in a counter-charge that sent the French fleeing into the river with heavy losses. This was the last battle in which a British sovereign led his troops in person.

> Now boys! Now for the honour of England! Fire, and behave brave and the French will run!
>
> **George II**, addressing his troops at the Battle of Dettingen.

Dien Bien Phu, Battle of

Decisive battle in the Indochina War between Vietnamese and French troops at a French fortress in north Vietnam. French troops were besieged from 13 March to 7 May 1954 by the communist Vietminh, and the eventual fall of Dien Bien Phu resulted in the end of French control of Indochina.

- The French commander, Gen Henri Navarre, hoped to draw the Vietminh into a pitched battle.
- He made no attempt to occupy the hills surrounding the French base.
- The Vietminh commander, Gen Vo Nguyen Giap, demolished the French strongholds one by one.
- The French could only send in supplies by parachute and many fell into Vietminh hands.
- A Vietminh siege was followed by a massive attack on 1 May 1954, and the French commander and 11,000 troops surrendered on 7 May.

Dieppe Raid

In World War II, a disastrous Allied attack of August 1942, on the German-held seaport on the English Channel northwest of Paris. Some 5,000 Canadian troops and 1,000 commandos took part, but found German defenders well prepared. The raid cost the Allies heavily, although valuable lessons about landing on hostile beaches were learned and applied in the D-Day landings of 1944.

At Dieppe, the Canadians lost 215 officers and 3,164 troops, the commandos 24 officers and 223 troops, the Royal Navy 81 officers, 469 sailors, and 34 ships, and the RAF lost 107 aircraft. In contrast, the Germans lost only 345 soldiers.

disarmament

Reduction of a country's weapons of war. Most disarmament talks since

World War II have been concerned with **nuclear warfare**, but **biological, chemical**, and conventional weapons have also come under discussion. Attempts to limit the arms race have included the Strategic Arms Limitation Talks (SALT) of the 1970s and the Strategic Arms Reduction Talks (START) of the 1980s–90s. The agreements on nuclear weapons were facilitated by the fear of nuclear war, a desire to limit military expenditure, and by technological developments, but the control of conventional armaments has proved more difficult. The UN Conference on Disarmament was set up in 1978 and is based in Geneva.

division
Military formation consisting of two or more **brigades**. A major general at divisional headquarters commands the brigades and also additional artillery, engineers, attack helicopters, and other logistic support. There are 10,000 or more soldiers in a division. Two or more divisions form a **corps**.

Dogger Bank, Battle of
World War I naval engagement between British and German forces that met by accident on 24 January 1915 at Dogger Bank, a sandbank in the middle of the North Sea. British casualties amounted to 6 killed and 22 wounded; German casualties came to about 1,000 killed and 300 wounded. The German battle cruiser *Blücher* was sunk, while the British flagship *Lion* was hit in the engine-room and halted. Admiral David **Beatty** transferred to a destroyer, but when the German fleet approached Heligoland the British disengaged due to the danger of attack by **submarines** and minefields.

Dönitz, Karl (1891–1980)
German admiral, originator of the 'wolf-pack' **submarine** technique, that sank 15 million tonnes of Allied shipping during the course of **World War II**. He was in charge of Germany's **U-boat** force until succeeding Raeder as commander-in-chief of the navy in January 1943 and devoting himself to trying to overcome Allied naval superiority. He succeeded **Hitler** as Germany's leader in 1945, capitulated, and was imprisoned 1946–56.

Dowding, Hugh Caswall Tremenheere, 1st Baron Dowding (1882–1970)
British air chief marshal, chief of Fighter Command at the outbreak of **World War II** in 1939, a post he held through the **Battle of Britain** in 1940. His refusal to commit more **fighters** to France earlier in 1940 proved to be a

vital factor in the later Battle of Britain, but his uncompromising attitude upset political leaders and he was replaced in November 1940. He retired in 1942.

dragoon
Mounted soldier who carried an infantry weapon such as a 'dragoon', or short **musket**, as used by the French army in the 16th century. Later regiments retained the name, even after the original meaning became obsolete.

dreadnought
Class of **battleships** built for the British navy after 1905 and far superior in speed and armaments to any other warship then afloat. The first dreadnought was launched in February 1906 with armaments consisting entirely of big guns. The German and other navies quickly followed with similar battleships.

Drake, Francis (c. 1540–1596)
English buccaneer and explorer. Having enriched himself as a pirate against Spanish interests in the Caribbean in 1567–72, he sailed around the world in 1577–80 in the *Golden Hind*, robbing Spanish ships as he went. This was the second circumnavigation of the globe. In a raid on Cádiz in 1587 Drake burned some 10,000 tonnes of shipping, 'singed the King of Spain's beard' (attributed remark), and delayed the invasion by the **Spanish Armada** for a year. Drake helped to defeat the Armada in 1588. He died of dysentery in Panama in 1596.

> ❦ There is plenty of time to win this game [of bowls], and to thrash the Spaniards too. ❧
>
> Famous remark, attributed to **Francis Drake.**

drill
In military usage, the repetition of certain fixed movements in response to set commands. It became important in the 18th century when infantry marched into battle in lines, or ranks, firing together on command. Drill is used in training, to improve discipline, and helps to get a body of soldiers

from one place to another in an orderly fashion, especially for parades and ceremonial purposes.

DUKW

US amphibious truck of **World War II**, basically a 2.5-ton cargo truck fitted with buoyancy tanks and with screw propulsion when in the water. Principally used to ferry supplies and men from ship to shore, the DUKW played a vital part in almost every amphibious landing performed by Allied troops, such as that of **D-Day**.

Dunkirk

French Dunkerque, most northerly seaport of France, at the entrance to the Strait of Dover, the scene in **World War II** of the evacuation of Allied troops after the German invasion of France 1940. In all, 337,131 Allied troops (including about 110,000 French) were evacuated from the beaches as German forces approached. The sea-borne evacuation of these troops in May–June 1940 (known as Operation Dynamo) was achieved by a motley 'fleet' of over 1,000 ships, from warships down to private yachts. It was a morale-boosting operation after the German onslaught and the fall of the Low Countries and France.

Easter Rising

An Irish republican insurrection against the British government that began on Easter Monday, April 1916, in Dublin. It is an important landmark in Irish history, though a military failure. The rising was organized by the Irish Republican Brotherhood (IRB), led by Patrick Pearse, along with sections of the Irish Volunteers and James Connolly's socialist Irish Citizen Army. Around 1,600 rebels occupied prominent buildings around central Dublin. British forces, under Gen Sir John Maxwell, shelled the rebel positions, killing numerous civilians, and after six days of fighting the rebels surrendered.

> ❦ In the name of God, and of the dead generations from which she receives her old traditions of nationhood, Ireland through us summons her children to her flag and strikes for freedom. ❧
>
> **Patrick Henry Pearse** speaking at the General Post Office, Dublin, during the Easter Rising 1916

Eben Emael

In **World War II**, daring assault on 10 May 1940 by German glider troops to capture a Belgian fort, Eben Emael, strategically placed at the junction of the Albert Canal and Maas River, north of Liège. A squad of 85 German glider troops landed on top of the fort and put the gun turrets out of action, but were unable to get into the fort itself. The following day German troops crossed the canal by boat to relieve the glider force, and the fort surrendered.

Ebro, Battle of

Principal battle of the Spanish Civil War (see **Civil War, Spanish**), between 24 July and 18 November 1938, in the vicinity of Gandesa, about 40 km/25 mi south of Lerida. By the time the battle ended on 18 November, the Republicans had lost about 30,000 dead, 20,000 wounded, and 20,000 prisoners, while the Nationalists lost 33,000 killed and wounded. This defeat effectively destroyed the International Brigade (volunteers aiding the Republican cause) and put an end to any hope of Republican victory.

eighty-eight

In **World War II**, nickname for German 88-mm/3.4-in anti-aircraft guns, later used as anti-tank guns. Although the gun was first tried as an anti-tank weapon during the Spanish **Civil War** in 1936, its use in this role was only really proved effective by **Rommel** when he used them to fend off British **tanks** attacking **Arras** in May 1940. He went on to use eighty-eights to good effect in the North African campaign, and the gun acquired its formidable reputation as a tank destroyer from then on.

Eisenhower, Dwight David ('Ike') (1890–1969)

Military leader and 34th president of the USA from 1953 to1960. A general in **World War II**, he commanded the Allied forces in Italy in 1943, then the Allied invasion of Europe, and from October 1944 all the Allied armies in the West. He was an able organizer and his good-natured temperament helped ease tensions between nationalities and commanders in the Allied forces.

Eisenhower was born at Denison, Texas. A graduate of West Point military academy in 1915, he served in a variety of army posts before World War II. After the war he commanded the US Occupation Forces in Germany, then returned to the USA to become Chief of Staff. He resigned from the army in 1952 to campaign for the presidency; he was elected by the Republicans, and re-elected by a wide margin in 1956.

> ❝ The eyes of the world are upon you. The hopes and prayers of liberty-loving people everywhere march with you. ❞
>
> **Dwight Eisenhower**, in an order to Allied troops, D-Day 1944.

El Alamein
Site in the northern Egyptian desert of two battles in **World War II**; *see* **Alamein, El, Battles of**.

electronic countermeasures (ECM)
Jamming or otherwise rendering useless an opponent's **radar**, radio, television, or other forms of telecommunication. This is important in war (domination of the electronic spectrum was a major factor in the Allied victory in the 1991 **Gulf War**) but jamming of radio and television transmissions also took place during the **Cold War**, especially by the East bloc.

enhanced-radiation weapon (ERW)
Another name for the neutron bomb, a small **H-bomb** (hydrogen bomb) for battlefield use. The neutron or ERW bomb has relatively high radiation but relatively low blast, designed to kill, using brief neutron radiation that leaves buildings and weaponry intact.
See also: *nuclear warfare*.

Enigma
German enciphering machine of **World War II** and, by extension, the codes generated by it. The British cracked the code in the spring of 1940 and the Allies gained much useful intelligence as the Germans believed the code unbreakable.

The machine resembled an electric typewriter; pressing a key sent a current through three rotors, the starting positions of which could be altered and 'stepped' at each keystroke. This meant that hitting the same key would not produce the same enciphered letter, making the resulting message proof against conventional code-breaking techniques.

English Civil War *see* **Civil War, English**.

Eugène, Prince of Savoy (1663–1736)
Full name François Eugène de Savoie Carignan, Austrian general who had many victories against the Turkish invaders (whom he expelled from Hungary in 1697 at the Battle of Zenta) and against France in the War of the Spanish Succession (he fought at the battles of **Blenheim**, Oudenaarde, and **Malplaquet**). He again defeated the Turks 1716–18, and fought a last campaign against the French 1734–35.

evacuation

Removal of civilian inhabitants from an area liable to ground attack, **air raids**, or other hazards to safer surroundings. The term is also applied to military evacuation. People who have been evacuated are known as evacuees.

Large-scale evacuation of civilians took place during **World War II** in the UK, when the government encouraged parents to send their children away from urban and industrial areas to the countryside in order to escape the threat of air raids. Allied troops were evacuated from the beaches of **Dunkirk** in 1940.

explosive

Any material capable of a sudden release of energy and the rapid formation of a large volume of gas, leading, when compressed, to the development of a high-pressure wave (blast). Many explosives (called low explosives) are capable of undergoing relatively slow combustion under suitable conditions. High explosives produce uncontrollable blasts. The first low explosive was **gunpowder**; the first high explosive was nitroglycerine.

Modern military explosives are often based on cyclonite (also called RDX). Plastic explosives, such as **Semtex**, are based on RDX mixed with oils and waxes. The explosive force of **atomic** and **hydrogen bombs** arises from the conversion of matter to energy.

Falaise Gap

Action during **World War II** at Falaise, a town southeast of Caen, in Normandy, northern France. In August 1944, some 50,000 German troops were captured by the Allies in the Falaise pocket as they attempted to avoid encirclement by British and Canadian troops advancing from the south and US troops from the east and north. The town was largely destroyed in the fighting and was rebuilt in the 1950s.

Falkland Islands, Battle of the

In **World War I**, a British naval victory (under Admiral Sir Frederick Sturdee) over German forces under Admiral Maximilian von Spee on 8 December 1914. Von Spee intended to bombard the Falklands before proceeding around the Cape of Good Hope to arouse the disaffected Boers of South Africa. However, a British force was stationed off the Falklands and when von Spee realized he had run into a trap he fled the area. The British gave chase and in the ensuing battle von Spee's squadron was entirely destroyed with a loss of 2,100 crew.

Falklands War

War between Argentina and Britain over disputed sovereignty of the Falkland Islands initiated when Argentina invaded and occupied the islands

> ❝ The British soldiers didn't look like men who had just walked across the island but they had, every step of the way on their own two feet. Fifty miles they'd come over mountains and bogs in weather that chilled the bone and soaked the skin, and at the end of it they'd fought bravely and well. ❞
>
> Journalist **Brian Hanrahan**, reporting the *Falklands War* for BBC TV news, 25 June 1982, (recorded in *'I Counted Them All Out and I Counted Them All Back',* 1982)

on 2 April 1982. On the following day, the United Nations Security Council passed a resolution calling for Argentina to withdraw. A British task force was immediately dispatched to the South Atlantic and, after a fierce conflict in which more than 1,000 Argentine and British lives were lost, 12,000 Argentine troops surrendered and the islands were returned to British rule on 14–15 June 1982.

fallout
Harmful radioactive material released into the atmosphere in the debris of a nuclear explosion (see **nuclear warfare**) and descending to the surface of the Earth. Such material can enter the food chain, cause radiation sickness, and last for hundreds of thousands of years.

field marshal
The highest rank in many European armies. A British field marshal is equivalent to a US **general** (of the army). George II introduced the rank to Britain from Germany in 1736.

fifth column
Group within a country secretly aiding an enemy attacking from without. The term originated 1936 during the Spanish **Civil War**, when Gen Mola boasted that Franco's supporters were attacking Madrid with four columns and that they had a 'fifth column' inside the city.

fighter plane
Aeroplane used in warfare primarily to attack enemy aircraft, particularly **bombers,** but also used for reconnaissance and ground attacks. The type originated in the 'scouts' used in **World War I**, which met one another in 'dog-fights'. By **World War II**, monoplane fighters armed with machine guns and cannon had been developed. They included the British Spitfire, German Me 109, and US Mustang. The first jet fighter to see action was the German Me 262 in the closing months of the war. Modern fighters fly at two or three times the speed of sound, and carry missiles and bombs, as well as sophisticated electronics equipment.

firearm
Weapon from which projectiles are discharged by the combustion of an **explosive**. Firearms are generally divided into **artillery** (ordnance or **cannon**), with a bore greater than 2.54 cm/1 in, and small arms such as

rifles and **revolvers**, with a bore of less than 2.54 cm/1 in.

Although **gunpowder** was known in Europe 60 years previously, the invention of the gun dates from around 1300 to 1325, and is attributed to Berthold Schwartz, a German monk.

firestorm

Fire which rapidly grows out of control by sucking in surrounding air and thus feeding itself into an accelerating cycle. The incoming air soon reaches very high speeds, sufficient to knock people over or even sweep them into the fire, and the fires themselves become uncontrollable, causing immense devastation in urban areas. Firestorms are associated with intense incendiary bombing **air raids**; possibly the most notorious example was the Allied bombing raid on Dresden in February 1945, although the effect was first seen in the RAF raid on Hamburg in July 1943.

First World War

Another name for **World War I**, 1914–18.

Fisher, John Arbuthnot, 1st Baron Fisher (1841–1920)

British admiral, UK First Sea Lord 1904–10, who carried out many radical reforms and innovations, including the introduction of the dreadnought **battleship**. He served in the **Crimean War** in 1855 and the China War 1859–60. He held various commands before becoming First Sea Lord, and returned to the post in 1914, but resigned the following year, disagreeing with Winston **Churchill** over sending more ships to the **Dardanelles** in World War I.

flag

Piece of cloth used as an emblem or symbol for nationalistic, religious, or military displays, or as a means of signalling. In warfare, flags have been important as rallying points, and long used for unit identification and message-sending.

Ancient armies had flags and other forms of standard, carried into battle. The first Roman flag was apparently the *vexillum*, the standard of the cavalry. One of the earliest medieval flags was the gonfalon, a square or oblong

---

The union flag was ordered by James I to be borne at the maintop of all British ships except ships of war, which bore it on the jackstaff at the end of the bowsprit, which is why it is, (incorrectly), known as the Union Jack.

---

piece of cloth, sometimes with streamers, attached to a bar or frame. At first each company of an English regiment had its distinctive flag or colour, but in the early 1700s the number of flags in each regiment was reduced to two, the royal and regimental colours.

flak

Term used by Allied troops and airmen in **World War II** to describe anti-aircraft fire from the German abbreviation for *Flugzeugabwehrkanone* – 'aircraft attack gun'. A flak jacket was an **armoured** jacket worn by air crews as a defence against bullets and shell fragments.

flamethrower

Weapon emitting a stream of burning liquid that can be directed against troops or strongholds. Flamethrowers were first used by the Germans in **World War I** at the Battle of Hooge, July 1915. The weapon consisted of a back-pack with a reservoir of compressed nitrogen and a tank containing a 'flame liquid', usually a mixture of coal-tar and benzine. The gas pressure was sufficient to give the flaming liquid a range of about 45 m/50 yds.

Flanders, Battle of

In **World War I**, the series of actions as the British troops advanced into Belgium and northern France during September – November 1918, driving the German forces out of the Benelux area and back into Germany. This encompasses the battles of the Yser and **Ypres**. The British commander, Field Marshal Douglas Haig, conceived the overall plan for what was to be the last major campaign of the war.

Flodden, Battle of

Defeat of the Scots by the English under the Earl of Surrey on 9 September 1513, on a site 5 km/3 mi southeast of Coldstream, in Northumberland, England. James IV of Scotland, declaring himself the active ally of France, crossed the border into England with an invading army of 30,000. The Scots were defeated, suffering heavy losses, and James himself was killed.

Flying Tigers

Nickname given to the American Volunteer Group in **World War II**, a group of US pilots recruited to fight in China by Maj-Gen Chennault 1940–41. The group proved an effective force against the Japanese over southern

China and Burma in 1941–42, destroying some 300 enemy aircraft. It was absorbed into the regular US air forces as the 14th Air Force with Chennault as its commander in 1942.

Foreign Legion

Volunteer corps of foreigners within a country's army. The French *Légion Etrangère*, founded in 1831, is one of a number of such forces. Enlisted volunteers are of any nationality (about half are now French), but the officers are usually French. Headquarters until 1962 was in Sidi Bel Abbés, Algeria; the main base is now Corsica, with reception headquarters at Aubagne, near Marseille, France.

fortification

Military defensive structure, originally a natural defence such as a hilltop, fortified by digging earthbanks and erecting fences as barricades. In Iron Age Europe, this led to the construction of large **hillforts**. Later, the Roman fort and the stone **castle** of medieval times were army bases as well as strongholds. The invention of **cannon** made stone walls vulnerable, and later fortifications employed a combination of walls, trenches, towers, and gun-positions. In **World War I**, both sides dug vast networks of trenches with barbed wire entanglements and minefields to deter the enemy from advancing.
See also: *Maginot Line; Vauban.*

Franco-Prussian War

War of 1870–71 in which Prussia defeated France. The Prussian chancellor Otto von **Bismarck** put forward a German candidate for the vacant Spanish throne with the deliberate, and successful, intention of provoking the French emperor Napoleon III into declaring war. The Prussians defeated the French at **Sedan**, then besieged Paris. The Treaty of Frankfurt of May 1871 gave Alsace Lorraine, and a large French indemnity to Prussia. The war established Prussia, at the head of a newly established German empire, as Europe's leading power.

Frederick (II) the Great (1712–1786)

King of Prussia from 1740, when he succeeded his father Frederick William, and a renowned military commander. In 1740 he started the War of the **Austrian Succession** by his attack on Austria. Through the peace of 1745 he secured Silesia. The struggle was renewed in the **Seven Years' War**, 1756–63, when Frederick had a hard struggle against the Austrians and their Russian allies; his success proved him to

> Frederick received a harsh military education from his father, and in 1730 was threatened with death for attempting to run away.

be one of the great soldiers of history. He acquired West Prussia in the first partition of Poland in 1772 and left Prussia as Germany's foremost state.

Fredericksburg, Battle of

In the American Civil War (*see* **Civil War, American**), a Confederate victory on 11–15 December 1862 over Union forces on the Rapahannock River close to Fredericksburg, Virginia. Although the Confederates led by **Lee** halted the Union march on Richmond, losses on both sides were heavy: The Union force numbered 125,000 troops under Gen Ambrose Burnside. Lee had some 85,000 troops. Union casualties were 13,000 dead and wounded; Confederate casualties 5,000, although many were only lightly wounded.

French Revolutionary Wars

Wars that began in 1792 between revolutionary France and other European nations anxious to restore the status quo. Austria attacked in April 1792, following the imprisonment of the French king Louis XVI but the French victory at Valmy saved the revolution. The Revolutionary Wars produced a brilliant campaigner in **Napoleon** Bonaparte, whose victories helped him seize control of the government in 1799. The wars are thenceforth known as the **Napoleonic Wars**.

frigate

Escort warship smaller than a **destroyer**. In the 18th and 19th centuries a frigate was a small, fast sailing warship. Before 1975 the term referred to a warship larger than a destroyer but smaller than a light cruiser. Today the frigate is the most numerous type of large surface vessel in the British **Royal Navy**. Britain's type-23 frigate (1988) is armoured, heavily armed (4.5-in/114-mm naval guns, 32 Sea Wolf anti-missiles and anti-aircraft missiles, and a surface-to-surface missile), and, for locating submarines, has a helicopter and a hydrophone array towed astern. Engines are diesel-electric up to 17 knots (31.5 kph), with gas turbines for spurts of speed to 28 knots (52 kph).

frogman

Underwater warfare diver; the name derives from the diver's appearance when wearing wetsuits, flippers, and masks. Used in **World War II** and subsequently, frogmen are usually associated with sabotage and attacks on shipping, but also carry out reconnaissance of beaches and underwater obstacles before an amphibious operation.

Galland, Adolf (1912–1996)
German air ace of **World War II**. He served in the **Spanish Civil War** of 1936–39 and then the Polish and French campaigns of 1939 and the **Battle of Britain** 1940. He was promoted to general in charge of fighter operations on all European fronts in November 1941 but was relieved of his post late in 1944, and at **Hitler's** personal request trained and led a **jet** fighter squadron. He raised his score to 104 enemy aircraft before injuries put an end to his flying career.

Gallipoli
Port in European Turkey, on the Dardanelles, giving its name to the peninsula (ancient name Chersonesus) on which it stands. In **World War I**, at the instigation of Winston **Churchill**, an unsuccessful attempt was made from February 1915 to January 1916 by Allied troops to force their way through the **Dardanelles** and link up with Russia. The campaign was fought mainly by Australian and New Zealand (Anzac) forces, who suffered heavy losses. An estimated 36,000 Commonwealth troops died during the campaign.

Garand rifle
US rifle M1, standard US military **rifle** from 1936 until the late 1950s. The first automatic rifle to be adopted as standard by any major power, its success in combat led to a reappraisal of this type of weapon by all combatants.

The Garand was a .30 in calibre, gas-operated, semi-automatic weapon firing from an eight shot magazine; it was replaced by the M14, a modified design, 7.62 mm/0.3 in NATO calibre, using a 20-round magazine.

Garibaldi, Giuseppe (1807–1882)
Italian soldier who played a central role in the unification of Italy. A patriot condemned to death for treason, he had to leave Italy and became a **mercenary** soldier while in exile in South America. He returned to fight in the 1848 revolution and subsequently lived on the island of Caprera. In

> ❝ I offer you hunger, thirst, forced marches, battles, and death. Anyone who loves his country, follow me. ❞
>
> **Garibaldi**

1860 he led his force of 1,000 redshirts to victory, conquering Sicily and Naples. He later fought in the 1866 Austrian War and in the **Franco-Prussian War** (1870–71), for France.

Garros, Roland (1888–1918)

French fighter pilot in World War I. He held several aviation records from 1911–1912. He had some success as a fighter pilot but crash-landed in enemy territory in April 1915 and was taken prisoner. He escaped in February 1918 and returned to France, rejoined the air service and scored several more victories before being shot down and killed on 5 October 1918.

Garibaldi, Giuseppe *Garibaldi, Italian patriot who unified Italy.*

Garros was the first pilot to fire **machine guns** through the propeller blades. He had sheets of steel attached to the propeller blades to deflect bullets.

gas mask

Face mask designed to filter out or neutralize chemical agents from the air inhaled by the wearer. The standard design consists of a moulded face mask with goggles and a canister connected to the mask by a flexible pipe. The first gas masks, produced in **World War I**, were simply cotton pads soaked in various chemicals to protect against chlorine gas. Then came the 'gas helmet', a flannel bag soaked in chemicals and with a celluloid window in the front. Breathing drew air through the saturated flannel, neutralizing the gas. Exhaled air was expelled through the bottom of the bag suspended around the soldier's neck. Gas masks were issued to civilians during **World War II**, but the expected gas attacks from the air did not materialize.

gas warfare
Military use of gas to produce a toxic effect on the human body.
See also: *chemical warfare*.

Gatling, Richard Jordan (1818–1903)
US inventor of a rapid-fire gun. Patented in 1862, the Gatling gun had ten barrels arranged as a cylinder rotated by a hand crank. Cartridges from an overhead hopper or drum dropped into the breech mechanism, which loaded, fired, and extracted them at a rate of 320 rounds per minute.

The Gatling gun was used in the American **Civil War**, in the Indian Wars that followed the settling of the American West, and in the **Franco-Prussian War** of 1870. By 1882 rates of fire of up to 1,200 rounds per minute were achieved, but the weapon was soon superseded by Hiram Maxim's **machine gun** in 1889.

Gaugamela, Battle of
Also called the Battle of Arbela, decisive defeat in 331 BC of the Persians under Darius III (ruled about 380–330 BC) by the Macedonian king **Alexander the Great**. Alexander's tactical superiority enabled his 47,000 troops to defeat the massed Persian army of about 120,000 and the defeated Darius fled. He was later killed by his own troops, effectively giving Alexander the whole of Asia Minor.

The battle took place at the ancient town of Gaugamela, on the east bank of the River Tigris northeast of Nineveh (now Al Mawsil, Iraq).

general
Senior military rank, the ascending grades being major general, lieutenant general, and general. The US rank of general of the army is equivalent to the British **field marshal**.

Geneva Convention
International agreement of 1864 regulating the treatment of those wounded in war, and later extended to cover the types of weapons allowed, the treatment of prisoners and the sick, and the protection of civilians in wartime. The rules were revised at conventions held in 1906, 1929, and 1949, and by the 1977 Additional Protocols.

Genghis Khan (c. 1155–1227)
Mongol conqueror, ruler of all Mongol peoples from 1206. He conquered the empires of northern China 1211–15 and Khwarazm 1219–21, and invaded northern India in 1221, while his lieutenants advanced as far as the Crimea. Masters of cavalry tactics and merciless in war, the Mongols were invincible under his command

VAST EMPIRE

- Genghis Khan's empire ranged from the Yellow Sea to the Black Sea.
- He controlled probably a larger area than any other individual in history.
- His empire continued to expand after his death until it extended from Hungary to Korea.

Genghis Khan *Ruler of the world's largest-ever empire.*

genocide

Deliberate and systematic destruction of a national, racial, religious, or ethnic group defined by the exterminators as undesirable. The term is commonly applied to the policies of the Nazis during **World War II** (what they termed the 'final solution' – the extermination of all 'undesirables' in occupied Europe, particularly the Jews, *see* **Holocaust**). It is also referred to as 'ethnic cleansing'.

In 1948 the United Nations General Assembly adopted a convention on the prevention and punishment of genocide, as well as the Universal Declaration of Human Rights.

Gestapo

Contraction of *Geheime Staatspolizei*, Nazi Germany's secret police. The Gestapo was created by Hermann Goering to replace the political police and was transferred to the control of Heinrich Himmler's **SS** in 1934, under the command of Reinhard Heydrich. The Gestapo had sweeping powers and was one of the most feared and brutal elements of the **Nazi** regime, using torture and terrorism to stamp out anti-Nazi resistance. It was declared a criminal organization at the Nuremberg Trials 1946.

Gettysburg

One of the decisive battles of the American **Civil War**: a Confederate defeat by Union forces on 1–3 July 1863, at Gettysburg, Pennsylvania, 80 km/50 mi northwest of Baltimore. The South's heavy losses at Gettysburg came in the same week as their defeat at **Vicksburg**, and the Confederacy remained on the defensive for the rest of the war. The battle ended Robert E **Lee's** invasion of the North.

> **GETTYSBURG ADDRESS**
> - The battlefield site is now a national cemetery, at the dedication of which President Lincoln delivered the Gettysburg Address on 19 November 1863.
> - In this speech he reiterated the principles of freedom, equality, and democracy embodied in the US Constitution.

GI

Abbreviation for 'government issue', hence (in the USA) a common soldier.

Gibraltar, siege of

During the **American Revolution**, an unsuccessful Franco–Spanish blockade of the British-held fortress of Gibraltar, between June 1779 and February 1783. The **siege** inflicted great hardship: few supply ships were able to run the blockade and the residents came close to starvation. Throughout the siege, there were regular **artillery** exchanges. The final Spanish attack took place in September 1872 with an army of 40,000, aided by the combined French and Spanish fleets. Great losses were inflicted on the attacking force and the siege was finally raised on 6 February 1783.

'Glorious First of June'

Naval battle between the British and French fleets off Ushant on 1 June 1794, the first major naval action of the **French Revolutionary Wars**. Both sides claimed a victory; the British because they had damaged the French fleet and there was no major naval action for the next two years; the French because their fleet, damaged as it was, was not destroyed, and the British failed to prevent a vital grain convoy reaching France.

Goose Green

In the **Falklands War**, British victory over Argentina at Goose Green, south of San Carlos, on 28 May 1982 during the advance on Port Stanley.

British troops landed at and around Port San Carlos on the western side of the West Falkland island on 21 May and prepared to advance to Port Stanley, some 80 km/50 mi to the east. An Argentine force at Goose Green, to the south of San Carlos, posed a threat to the flank of the British advance. Troops of the Parachute Regiment made an attack with limited support. The Argentine garrison was overrun, and the survivors taken prisoner.

Gotha bomber

Early twin-engined biplane **bomber**, the principal strategic bomber of the German air force in **World War I** from 1917 onwards. Although early models (G1 and G2) were used on the Balkan and Eastern Fronts as tactical bombers they were unreliable and it was not until the improved G4 that Gotha became the standard heavy bomber for raids over England and France.

Grant, Ulysses S(impson) (1822–1885)

Born Hiram Ulysses Grant, American **Civil War** general and 18th president of the USA 1869–77. He had an unsuccessful career in the army and in business 1839–54. On the outbreak of the Civil War he received a commission on the Mississippi front. He took command there in 1862, and by his capture of **Vicksburg** in 1863 brought the Mississippi front under Northern control. In 1864 he was made Union commander-in-chief. He wore down **Lee's** resistance, and in 1865 received his surrender at Appomattox. Grant was elected president in 1868 and re-elected in 1872.

> I know only two tunes. One of them is 'Yankee Doodle' and the other isn't.
>
> **Ulysses S Grant**, quoted in W E Woodward, *Meet General Grant*.

Great Marianas Turkey Shoot
In **World War II**, air battle during the naval Battle of the **Philippine Sea** on 20 June 1944. A Japanese fleet of 6 **aircraft carriers** with 342 aircraft set out to trap a US fleet between itself and land-based aircraft from Guam. Aircraft from the 15 carriers accompanying the US force intercepted the Japanese air force and shot down over 300, only a handful reached the US fleet and caused little or no damage.

Great Northern War
War of 1700–21 between Sweden led by Charles XII and an alliance of Peter the Great's Russia, Denmark-Norway, and Saxony-Poland. The war ended Sweden's status as the dominant power in the Baltic. The main battle was **Poltava** in 1709, a Swedish defeat.

Great Wall of China
Continuous defensive wall stretching from western Gansu to the Gulf of Liaodong (2,250 km/1,450 mi). It was built under the Qin dynasty from 214 BC to prevent incursions by the Turkish and Mongol peoples and extended westwards by the Han dynasty and was once even longer. Some 8 m/25 ft high, it consists of a brick-faced wall of earth and stone, has a series of square watchtowers, and has been carefully restored.

> The Great Wall of China is so large that it can be seen from space.

Greek fire
Combustible material used in medieval warfare, especially by the Byzantines, against wooden ships or fortifications. Like a **flame-thrower**, it could be aimed, usually through a tube, and would explode on impact. The main ingredient was almost certainly naphtha, or crude oil.

Greek fire was possibly invented in the 7th century by the Egyptian architect Callinicus who fled from Syria to Greece. The Byzantine Empire

used it until its fall in 1453, and kept its recipe a closely guarded secret.
See also: *napalm*.

grenade

Small missile, containing an **explosive** or other charge, usually thrown (hand grenade) but sometimes fired from a **rifle**. Hand grenades are generally fitted with a time fuse of about four seconds: time for the grenade to reach the target but not enough for the enemy to pick it up and throw it back.

Grenades were known in the 15th century, but were obsolete by the 19th century. They were revived in 1905 during the **Russo-Japanese War**, and again when trench warfare began in **World War I**. The British Mills bomb and the French 'pineapple' grenade were ball-like objects easily thrown, while the German stick grenade carried the metal canister of explosive on a wooden handle.

Guadalcanal, Battle of

In **World War II**, important US operation in 1942–43 on the largest of the Solomon Islands. The US discovered the Japanese were building an airfield and landed marines to take the site August 1942. The Japanese sent reinforcements by sea to recapture the airfield and a series of bitter engagements took place. The naval operations began to dwarf those on the land they were supposedly supporting and both sides lost large amounts of ships and aircraft. The engagements on land and sea were inconclusive until the Japanese concluded such heavy naval losses could not be justified by one island and evacuated on 7 February 1943.

Guam

Battle site in **World War II** on the southernmost of the Mariana Islands in the West Pacific, now an unincorporated territory of the USA. Guam was occupied by Japan as an air and naval base 1941–44. The July 1944 battle in which Americans recaptured the island caused heavy damage. Guam became the headquarters of the US Pacific Strategic Air Command in 1954 and is now also the central command for all US naval operations in the West Pacific.

Guderian, Heinz Wilhelm (1888–1954)

German general in **World War II**. He created the **Panzer** (German 'armour') divisions that formed the ground spearhead of Hitler's **Blitzkrieg** attack strategy, achieving a significant breakthrough at Sedan in Ardennes, France, in 1940, and leading the advance to Moscow in 1941. Determined Soviet resistance led him to make a partial withdrawal and **Hitler** dismissed him from his post. He was reinstated as inspector general of armoured troops in 1943 and became Chief of Staff after the July 1944 plot against Hitler, but was again dismissed by Hitler in March 1945.

guerrilla
(Spanish 'little war') irregular soldier fighting in a small, unofficial unit, typically against an established or occupying power, and engaging in sabotage, ambush, and the like, rather than pitched battles against an opposing army. Guerrilla tactics have been used both by resistance armies in wartime (for example, the **Vietnam War**) and in peacetime by national liberation groups and militant political extremists (for example, the Sri Lankan Tamil Tigers).

The term was first applied to the Spanish and Portuguese resistance to French occupation during the **Peninsular War** (1808–14).

Significant guerrilla groups active in recent times include:

- Hamas (Islamic Resistance Movement), Palestinian Islamic fundamentalist organization, opposed to the Palestine Liberation Organization's 1993 peace accord with Israel.
- Hezbollah, Shiite Muslim militia organization in Lebanon.
- **Irish Republican Army (IRA)**, organization committed to the formation of a unified Irish republic.
- Palestine Liberation Organization (PLO), from 1993 pledged to peaceful coexistence with Israel.
- Tamil Tigers, Tamil separatist organization in Sri Lanka. A ceasefire broke down in 1995, and the organization was outlawed in 1998.

Guevara, Che (Ernesto) (1928–1967)
Marxist revolutionary, renowned for his **guerrilla** techniques. Born in Argentina, he trained as a doctor, but left his homeland in 1953 because of his opposition to President Juan Perón. In effecting the Cuban revolution of 1959 against the Cuban dictator Fulgencio Batista, he was second only to Fidel Castro and Castro's brother Raúl. In 1965 he went to the Congo to fight against white **mercenaries**, and then to Bolivia, where he was killed in

an unsuccessful attempt to lead a peasant rising near Vallegrande. In 1997 his remains were returned to Cuba for a hero's burial.

Gujarat

In the Punjab, India, scene of a battle in the Second Sikh War on 21 February 1849, when 60,000 Sikhs were defeated by British artillery, infantry, and cavalry under Lord Gough. As many as 50,000 Sikhs are thought to have died. The remainder of the Sikh armies surrendered on 10 March, ending the war.

> Gujarat was the first battle in which surgeons used anaesthetics in the field to carry out amputations.

Gulf War

War from 16 January to 28 February 1991 between Iraq and a coalition of 28 nations led by the USA. The invasion and annexation of Kuwait by Iraq on 2 August 1990 provoked a build-up of US troops in Saudi Arabia, eventually totalling over 500,000. The UK subsequently deployed 42,000 troops, France 15,000, Egypt 20,000, and other nations smaller contingents.

An air offensive lasting six weeks, in which 'smart' weapons came of age, destroyed about one-third of Iraqi equipment and inflicted massive casualties. A 100-hour ground war followed, which effectively destroyed the remnants of the 500,000-strong Iraqi army in or near Kuwait. Estimates of Iraqi casualties are in the range of 80,000–150,000 troops and 100,000–200,000 civilians.

- About 90,000 tonnes of ordnance was dropped by US planes on Iraq and occupied Kuwait
- of these, precision-guided weapons accounted for 7%
- of these, 90% hit their targets
- only 25% of the conventional bombs did so.

gunpowder

Also called black powder, the oldest known **explosive**, a mixture of 75% potassium nitrate (saltpetre), 15% charcoal, and 10% sulphur. It was first used by the Chinese in **rockets** and by Europeans in **cannon** during the Middle Ages.

Gurkha

Member of any of several peoples living in the mountains of Nepal, whose young men have been recruited since 1815 for the British and Indian armies. The ten regiments of Gurkhas served in both World Wars. When India and Pakistan became independent in 1947, four of the Gurkha regiments were assigned to the British army, and six to the Indian. Gurhkas still serve in the modern British army.

Gustavus Adolphus (1594–1632)

Gustavus II or Gustaf II, King of Sweden from 1611, when he succeeded his father Charles IX. He waged successful wars with Denmark, Russia, and Poland, and in the **Thirty Years' War** became a champion of the Protestant cause. Arriving in Germany in 1630, he defeated the German general **Wallenstein** at **Lützen**, southwest of Leipzig on 6 November 1632, but was killed in the battle. He was known as the 'Lion of the North'.

> ❝ I have taken the water from them; I would take the air if I could. ❞
>
> **Gustavus Adolphus** addressing Louis de Geer, who had tried to persuade him to lift the blockade of Gdansk during the Swedish–Polish conflict of 1627.

Haig, Douglas, 1st Earl Haig (1861–1928)
British army officer, commander-in-chief in **World War I**, born in Edinburgh, Scotland. His **Somme** offensive in France in the summer of 1916 made advances only at enormous cost to human life, and his **Passchendaele** offensive in Belgium from July to November 1917 achieved little at a similar loss. He was created **field marshal** in 1917 and, after retiring, became first president of the British Legion in 1921.

> ❝ Every position must be held to the last man: there must be no retirement. With our backs to the wall, and believing in the justice of our cause, each one must fight on to the end. ❞
>
> **Douglas Haig,** in an order, given on 12 April 1918.

Halsey, William Frederick (1882–1959)
US admiral. A highly skilled naval air tactician, his handling of **aircraft carrier** fleets in **World War II** played a significant role in the eventual defeat of Japan. He was appointed commander of US Task Force 16 in the Pacific in 1942 and almost immediately launched the Doolittle raid on Tokyo. He took part in operations throughout the Far East, including Santa Cruz, **Guadalcanal**, Bougainville, and the Battle of **Leyte Gulf**. He was promoted to fleet admiral in 1945 and retired in 1947.

Hampton Roads, Battle of
The first battle between armoured warships, known as ironclads, in the **American Civil War** between the Confederate *Merrimack* and the Union battleship *Monitor* on 8 March

In the battle, each captain watched their shots bounce off the other vessel. In the end *Monitor* ran out of ammunition, and *Merrimack* withdrew to make minor repairs.

1862 off the southeast coast of Virginia. The *Merrimack* had recently been renamed the *Virginia*. Neither vessel made any impression on the other after several hours of exchanging fire and eventually both withdrew. When the Union army later captured Norfolk, the Confederates set *Merrimack* on fire. *Monitor* was sunk in a storm while sailing back to New York on 31 December 1862.

hand grenade *see* **grenade**.

Hannibal, 'the Great' (247–182 BC)
Carthaginian general from 221 BC, son of Hamilcar Barca. His siege of Saguntum (now Sagunto, near Valencia) precipitated the Second **Punic War** with Rome. Following a campaign in Italy, Hannibal was the victor at Trasimene in 217 BC. He had an exceptional interest in military intelligence and reconnaissance, and his abilities as a tactician were shown at **Cannae** in 216 BC, but he failed to take Rome. In 203 BC he returned to Carthage to meet a Roman invasion but was defeated at **Zama** in 202 BC by Publius Cornelius Scipio (later Scipio 'Africanus'), and exiled in 196 BC. He eventually killed himself.

> Knowing that Rome controlled the seas, Hannibal decided to invade by the one route the Romans thought impossible – by land. In 218 he led his army across the Alps into the Po Valley – the celebrated march with elephants.

Harris, Arthur Travers (1892–1984)
British marshal of the **Royal Air Force** in **World War II**. Known as 'Bomber Harris', he was commander-in-chief of Bomber Command 1942–45. He was criticized for his policy of civilian-bombing selected cities in Germany; he authorized the fire-bombing raids on Dresden, in which more than 100,000 died. He also showed a flair for dramatic actions, such as the celebrated 'thousand bomber raid' on Cologne in May 1942. Harris was the only senior British commander not to receive a peerage after the war, and no medal was ever struck for the men of Bomber Command.

Hastings, Battle of
Battle on 14 October 1066 at which William, Duke of Normandy ('the Conqueror') defeated King Harold of England and took the throne. The site

is 10 km/6 mi inland from Hastings, at Senlac, Sussex; it is marked by Battle Abbey.

Having defeated an invasion by King Harald Hardrada of Norway at **Stamford Bridge**, Harold moved south with an army of 9,000 to counter the landing of the Duke of Normandy. The Normans dominated the battle with archers supported by cavalry, breaking through ranks of infantry. Both sides suffered heavy losses but Harold's death left England open to Norman rule.

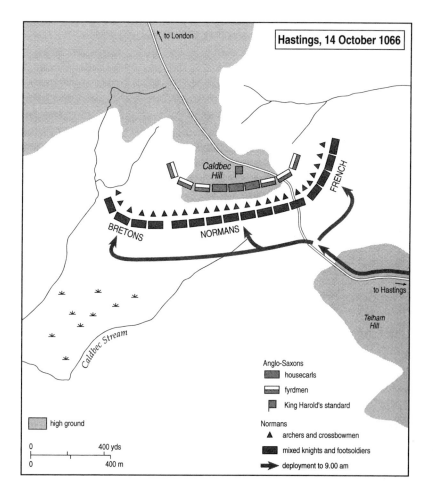

- At the turning point of the battle, some of Harold's personal guard broke ranks to pursue stragglers.
- William ordered part of his force to simulate flight.
- Many English troops ran down the hill after the Normans, who then turned and cut them down.
- William resumed his attack. With arrows falling about them, the English allowed the Norman foot soldiers to get among them.
- Harold and his two brothers were killed, and his army totally destroyed.

Hattin, Battle of

Crushing defeat of the **Crusaders** by **Saladin**, on 4 July 1187, at a village in Palestine, northwest of Tiberias. A column of Frankish Crusaders was attacked by a Saracen army and brought to a halt at Hattin, a place with no water. Harassing attacks during the night ensured that the Crusaders had no rest, and the lack of water demoralized them. The Saracens swept around in two wings and annihilated the Crusader force. The defeat destroyed the military power of the Kingdom of Jerusalem.

H-bomb (hydrogen bomb)

Bomb that works on the principle of nuclear fusion. A massive explosion results from the themonuclear release of energy when hydrogen nuclei are fused to form helium nuclei. The first hydrogen bomb was exploded at Eniwetok Atoll in the Pacific Ocean by the USA, 1952.
See also: *nuclear warfare*.

helicopter

Powered aircraft with a rotary wing, or rotor, that can take off and land vertically, move in any direction, or remain stationary in the air. Experiments using the concept of helicopter flight date from the early 1900s, with the first successful flight in 1907. Ukrainian-born US engineer Igor Sikorsky built the first practical single-rotor craft in the USA in 1939. In war, helicopters have been used extensively in **Korea** and **Vietnam**. They carry troops and equipment, make 'gunship' attacks on ground targets, ferry wounded to aid stations, and carry out air-sea rescues. At sea, ship-borne helicopters carry depth charges and homing **torpedoes** to attack **submarines** or surface targets.

Heligoland Bight, Battle of

World War I naval battle between British and German forces, on 28 August 1914, fought in the Heligoland Bight, the stretch of water between Heligoland Island and the German mainland used by the German fleet for exercises. The British launched a surprise raid on the German vessels exercising in the Bight and succeeded in sinking three light cruisers and a destroyer, a severe blow to German naval morale.

helmet

Protective head covering, often with a strap that goes under the chin. Metal helmets of various designs, including nose and cheek guards, were worn as part of a soldier's **armour** in ancient and medieval times. The steel helmet was reintroduced during **World War I** to protect soldiers from gunfire and flying shrapnel, and is still worn by combat troops.

helmet *Helmets from different periods from the Painted Chamber, Westminster.*

hill fort

European Iron Age site with massive banks and ditches for defence, used as both a military camp and a permanent settlement. Germanic peoples spread the tradition of forts with massive defences, timberwork reinforcements, and sometimes elaborately defended gateways, overlooked from a rampart walk, as at Maiden Castle in Dorset, England. Hill forts with a single ditch and bank usually date from the earliest part of the Iron Age, whereas those with more complex defences came later.
See also: *fortifications*.

Hindenburg Line

In **World War I**, German western line of **fortifications** running from Arras to Laon, built 1916–17. Part of the line was taken by the British in the third Battle of **Arras**, but it generally resisted attack until the British offensive of summer 1918. The fortifications were designed to allow the German Army to hold off the Anglo-French attack whilst dealing a decisive blow to Russia.

Hiroshima
Industrial city and port on the south coast of Honshu island, Japan. On 6 August 1945, towards the end of **World War II,** the city was utterly devastated by the first US atomic bomb dropped by the bomber *Enola Gay*. A second strike on Nagasaki followed three days later. More than 10 sq km/4 sq mi was obliterated, with very heavy damage outside that area. Casualties totalled at least 137,000 out of a population of 343,000, of these 78,150 were found dead, others died later. By the 1990s the death toll, which included individuals who died from radiation-related diseases, had climbed to 192,000. An annual commemorative ceremony is held on 6 August.

Hitler, Adolf (1889–1945)
Nazi dictator, born in Austria. He was Führer (leader) of the Nazi Party from 1921. He served as a volunteer in the German army during **World War I**. As dictator, Hitler conducted **World War II** in a ruthless but idiosyncratic way. He ordered his armies into neighbouring countries, imposed his rule through repressive occupation forces, and had millions of people killed in what became known as the **Holocaust**. He survived an assassination attempt by army officers in 1944, and killed himself during the battle for **Berlin** in the last days of the war.

Holland, John Philip (1840–1914)
Irish engineer who developed some of the first military **submarines** used by the US navy. He began work in Ireland in the late 1860s and later emigrated to the USA. His first successful submarine was launched in 1881 and, after several failures, he built the *Holland* in 1893, which was bought by the US navy in 1895. He introduced many of the innovations that would be incorporated in later attack submarines.

Holocaust
The annihiliation of more than 16 million people by the **Hitler** regime 1933–45 in extermination and concentration camps such as Auschwitz and Treblinka in Poland, and **Belsen**, Buchenwald, and Dachau in Germany. Of the victims, more than 6 million were Jews (over 67% of European Jewry); other persecuted groups included gypsies and the disabled. Victims were variously starved, tortured, experimented on, and worked to death. Many thousands were executed in gas chambers, shot, or hanged.

home front

Organized sectors of domestic activity in war time, mainly associated with **World Wars I** and **II**. Features of the UK home front in World War II included the organization of the **blackout**, **evacuation**, air-raid shelters, the Home Guard (a defence force of mainly over-age volunteers), rationing, and distribution of **gas masks**. With many men on active military service, women were called upon to carry out jobs previously undertaken only by men.

Hood

British battle cruiser of **World War II** sunk by gunfire from the German **battleship *Bismarck*** south of Greenland on 24 May 1941. Only three of the 1,420 crew survived. Of 41,900 tonnes/41,200 tons displacement and armed with eight 15-in (381-mm), twelve 5.5-in (140-mm), and eight 4-in (102-mm) guns, with four torpedo tubes, it could reach a speed of 31 knots (57 kph).

home front *World War II propaganda poster.*

hoplite

In ancient Greece a heavily armed **infantry** soldier. The hoplite sword, or ksiphos, was used from the 9th to the 3rd centuries BC, during the Greek classical period. It is arguably the most successful **sword** of the ancient world. It was basically a slashing weapon weighted towards the point, but many examples are long cut-and-thrust weapons.

howitzer

Cannon, in use since the 16th century, with a particularly steep angle of fire. It was much developed in **World War I** for demolishing the fortresses of the trench system. The modern NATO FH70 field howitzer is mobile and fires, under computer control, three 43-kg/95-lb shells at 32 km/20 mi range in 15 seconds.

human torpedo

Torpedo fitted with seats for two men, with steering controls, and with a detachable warhead which can be attached to a target ship. They were first used by the Italian Navy during **World War II** in December 1941 in Alexandria harbour. The British battleships *Queen Elizabeth* and *Valiant* were put out of action for several months. In a British attack in January 1943, on Palermo harbour, a newly launched cruiser and a transport were sunk, and British human torpedoes sank another cruiser in the harbour of La Spezia in June 1944.

Hundred Years' War

Series of conflicts between England and France from 1337 to 1453. Its origins lay with the English kings' possession of Gascony (southwest France), which the French kings claimed, and with trade rivalries over Flanders. English triumphs in the first phase of the war included the naval victory at Sluys in 1340, **Crécy** in 1346, and **Poitiers** in 1356. In the later phase, Henry V defeated the French at **Agincourt** in 1415, a battle again won by longbow tactics. Despite further English victories, the war went the way of France. **Joan of Arc** helped save Orléans and France regained military superiority, partly through their improvement of cannon. The French victory at Castillon in 1453 is seen as marking the end of the war.

> ❛.. the King of France, hardened in his malice, would assent to no peace or treaty, but called together his strong host to take into his hand the duchy of Aquitaine, declaring against all truth that it was forfeit to him. ❜
>
> **Edward III**, King of England from 1327, proclamation at the outbreak of the Hundred Years' War, 1337.

ICBM
Abbreviation for intercontinental ballistic missile. These rocket missiles have from 1968 been equipped with clusters of warheads (which can be directed to individual targets) and are known as multiple independently targetable re-entry vehicles, (**MIRVs**). The 1980s US-designed MX, (Peacekeeper), carries up to ten warheads in each missile. The submarine-launched Trident missiles are made by the USA. Each warhead has eight independently targetable re-entry vehicles (each nuclear-armed) with a range of about 6,400 km/4,000 mi and can fire on eight separate targets within about 240 km/150 mi of the central aiming point. The Trident system entered service within the Royal Navy in the 1990s.
See also: *missile; nuclear warfare.*

Immelmann, Max (1890–1916)
German **fighter** ace in **World War I**. He developed the 'Immelmann Turn', a manoeuvre in which, pursued, he would climb suddenly in a half-loop, roll, and then dive back at his pursuer. He was shot down and killed near Lens by Lt George McCubbin on 18 June 1916.

Imphal, Battle of
In **World War II**, Allied operation of 1944 to hold Japanese forces back from an important road junction in the Manipur district of northeast India, northwest of Calcutta; the turning point in the Burma campaign. Imphal was crucial to the Japanese plan for the invasion of India in 1944 and British general William **Slim** devoted three divisions and extensive air support to its defence. Imphal held out for three months, until the British were able to break the **siege**. The Japanese, starving and diseased, had lost 53,000 troops and fell back to the Chindwin River, abandoning their artillery and transport.

incendiary bomb
Bomb containing inflammable matter. Usually dropped by aircraft, incendiary bombs were used in **World War I** and incendiary shells were

used against Zeppelin **airships**. Incendiary bombs were a major weapon in attacks on cities in **World War II,** causing widespread destruction. To hinder firefighters, delayed-action high-explosive bombs were usually dropped with them. In the **Vietnam War**, US forces used **napalm** in incendiary bombs.

Inchon, Battle of

In the **Korean War**, successful US Marines amphibious operation on 15 September 1950 at Inchon, west of Seoul, South Korea. The Marines secured the city within two weeks and broke the North Korean forces' hold on the Pusan area.

infantry

Foot-soldier. In ancient times, foot-soldiers were the massed ranks of most armies, often untrained and ill-equipped. However, some infantry units, such as the Greek **hoplites** and Roman **legions**, were well disciplined and formidable. Before the development of **firearms**, infantry weapons included spears, swords, and pikes. Most archers also fought on foot. Infantry were **drilled** to march in lines or columns, and to form defensive formations to counter **cavalry**. From the 1600s, infantry were armed with **muskets** and in the 1800s with repeating **rifles**. In **World War I**, infantry troops fought bloody battles from trenches and fortifications, seldom advancing more than a few yards, and suffering enormous casualties. **World War II** saw the use of mobile infantry, operating alongside armoured units, a role they still perform in modern armed forces.

Inkerman, Battle of

In the **Crimean War**, British and French victory, on 5 November 1854, over Russians attacking the Inkerman Ridge, which was occupied by the British army besieging Sevastopol.

The Russians were able to seize the hilltop and bring up **artillery**. However, a rapid counterattack recaptured the hilltop. The Russians then launched a second attack, driven off by the arrival of British and French reinforcements. The fighting lasted about seven hours, during which time the British lost about 2,400 troops, the French about 1,000, and the Russians an estimated 11,000.

Iran–Iraq War

War between Iran and Iraq 1980–88, claimed by Iran to have begun with the Iraqi offensive of 21 September 1980, and by Iraq with the Iranian shelling of border posts on 4 September 1980. Occasioned by a boundary dispute over the Shatt-al-Arab waterway, it fundamentally arose because of Saddam Hussein's fear of Iran's encouragement of the Shiite majority in Iraq to rise against the Sunni government. An estimated one million people died in the war, which was marked by offensive and counter-offensive manouevres, using infantry, artillery, missiles, and tanks.

- Chemical weapons were used in the Iran-Iraq war.
- Cities and oil installations were the target for bombing raids and rocket attacks.
- 414 ships were hit by missiles, including, in 1987, the US frigate *Stark*, attacked by the Iraqi airforce.
- 90,000 of the soldiers that were killed were under 15.

Irish Republican Army (IRA)

Militant Irish nationalist organization formed in 1919, the paramilitary wing of Sinn Fein. Its attacks in Britain included an attempt to

Irish Republican Army *Early IRA volunteers at target practice.*

kill members of the UK cabinet during the 1984 Conservative Party conference in Brighton. It also attacked British military bases in the UK and in Europe.

> The IRA was founded in 1919 by Michael Collins as the successor to the Irish Volunteers of 1913.

Despite its close association with Sinn Fein, the IRA usually speaks as a separate organization. In 1969 the IRA split into two wings, 'official' and 'provisional'. The Provisional IRA, or Provos, carried on with terrorist activities. In July 1997 the IRA announced a ceasefire as part of the 'peace process'. Splinter groups such as 'Continuity IRA' and 'Real IRA' carried out further attacks, including a car bombing attack in Omagh, County Tyrone, in August 1998, which claimed more than 26 lives.

ironclad
Wooden warship covered with armour plate. The first to be constructed was the French *Gloire* in 1858, but the first to be launched was the British HMS *Warrior* in 1859. The first battle between ironclads took place during the American **Civil War**, when the Union *Monitor* fought the Confederate *Virginia* (formerly the *Merrimack*) on 9 March 1862 at **Hampton Roads**. The design was replaced by **battleships** of all-metal construction in the 1890s.

Iron Cross
Medal awarded for valour in the German armed forces. Instituted in Prussia in 1813, it consists of a Maltese cross of iron, edged with silver.

Isandhlwana, Battle of
In the Anglo–Zulu War, Zulu victory over British forces on 22 January 1879, approximately 160 km/100 mi north of Durban. Only about 350 troops of the original contingent of 1,800 escaped, and the invasion of Zululand was temporarily halted until reinforcements were received from Britain. The 21 officers and 534 soldiers of the 24th Regiment were killed where they fought; there were no wounded, no prisoners, and none missing.

> ❦ We have certainly been seriously under-rating the power of the Zulu army. ❧
>
> Remark attributed to **Lord Chelmsford**, British viceroy of India, after the British defeat at Isandhlwana.

Issus, Battle of

Battle in 333 BC in which **Alexander the Great** defeated the Persian king Darius III at the ancient port of Issus in Cilicia, about 80 km/50 mi west of present-day Adana, Turkey. Darius' family were captured during the battle which secured Alexander's supply route in preparation for his invasion of the Persian Empire.

Alexander, with an army of 35,000 Macedonians, launched his **cavalry** against the Persian cavalry and routed them. The Macedonian foot soldiers then crossed the river and assaulted the Persian centre, while Alexander personally led his own cavalry against Darius' bodyguard, who fled from the field.

Iwo Jima, Battle of

Intense fighting between Japanese and US forces between 19 February and 17 March 1945 during **World War II**. In February 1945, US **marines** landed on the island of Iwo Jima, a Japanese air base, intending to use it to prepare for a planned final assault on mainland Japan. The 22,000 Japanese troops put up a fanatical resistance but the island was finally secured on 16 March. US casualties came to 6,891 killed and 18,700 wounded, while only 212 of the Japanese garrison survived.

Jena, Battle of
Also known as the Battle of Auerstadt, in the **Napoleonic Wars**, comprehensive French victory over the combined Prussian and Saxon armies on 14 October 1806 at Jena, Germany, 90 km/56 mi southwest of Leipzig. Prussian and Saxon losses amounted to some 40,000 troops and 200 guns, while French casualties were around 14,000. Napoleon so broke the Prussian forces that they were unable to prevent him from marching on Berlin. This disaster led to the complete overhaul and reorganization of the Prussian Army, which laid the foundations for its subsequent military prowess.

Jervis Bay
Armed British merchant ship of **World War II**, sunk by the German **battleship** *Admiral Scheer* whilst protecting a **convoy** of merchant ships in the North Atlantic on November 1940.

The *Jervis Bay's* Captain Fegen ordered the convoy to scatter and then, hopelessly outgunned, began firing at the battleship so as to draw it away from the convoy. The end was inevitable, and the *Jervis Bay* went down fighting, but 32 of the convoy ships managed to escape. Fegen was posthumously awarded the Victoria Cross.

jet aircraft
Aircraft powered by a jet engine, a kind of gas turbine. Jet propulsion was developed in the 1930s, and first used in military aircraft during **World War II**, by the Germans who developed the **V-1** flying bomb and fighters such as the Me 262. Jet fighters were faster than any piston-engined, propeller-driven types, and had replaced such aircraft in most roles by the 1950s. Jet aircraft are now used by all major air forces, and range from supersonic strike aircraft to heavy transports and bombers.

Joan of Arc, St (c. 1412–1431)
Also known as Jeanne d'Arc, French military leader who inspired the French at **Orléans** in 1428–29 and at **Patay**, north of Orléans, in 1429. As a young

peasant girl, she was the wrong age, class, and gender to engage in warfare, yet her 'heavenly voices' instructed her to expel the occupying English from northern France during the **Hundred Years' War** and to secure the coronation of Charles VII of France. She was captured on May 1430 at Compiègne by the Burgundians, who sold her to the English. She was found guilty of witchcraft and heresy by a tribunal of French ecclesiastics who supported the English, and was burned to death at the stake in Rouen on 30 May 1431.

> When I have done that which I have been sent by God to do, then I shall put on women's clothes.
>
> **Joan of Arc**, statement during her trial, 1431.

Jutland, Battle of

World War I naval battle between British and German forces on 31 May 1916, off the west coast of Jutland. Its outcome was indecisive, but the German fleet remained in port for the rest of the war.

Early on 31 May the German fleet under Admiral Scheer entered the North Sea from the Baltic. Before the main fleets met, a long range gunnery duel took place in which the British lost battle cruisers and sustained damage to their flagship. In the general melee that ensued another British battle-cruiser was sunk. However, the Germans realized they were outgunned and fled. The British commander Jellicoe, fearful of torpedoes in the failing light of evening, decided not to follow and the battle thus came to a somewhat inconclusive end.

kamikaze
Japanese 'wind of the gods', a name for pilots of the Japanese air force in **World War II** who deliberately crash-dived their planes, loaded with bombs, usually on to ships of the US Navy.

A special force of these suicide pilots was established in 1944. Initially many different types of aircraft were used but later the Okha, a specifically designed, piloted flying bomb, was developed. Kamikaze squads caused major problems for Allied shipping, sinking or severely damaging at least six major vessels from November 1944 to January 1945, until their base in the Philippines was destroyed by Allied air strikes.

Kasserine Pass
Pass in the Memetcha Mountains west of Gafsa, Tunisia, of great strategic importance in the North African campaign in **World War II**.

Held by US II Corps under Gen Fredendall, it was attacked by the German 5th **Panzer** Army on 14 February 1943. The US forces were shattered, as were two US counterattacks on the following day. Gen **Alexander** quickly moved British and US forces into blocking positions to halt the German thrust before it could get clear of the mountains, but **Rommel** withdrew so skilfully that the Allies never realized he was gone. Gen Fredendall was subsequently relieved of his command and replaced by Gen **Patton**.

Katyusha
Also called Stalin's Organ, a Soviet free-flight **rocket** of **World War II**. It was fired from racks mounted on a heavy truck and had a range of about 5 km/3 mi. Each truck had 48 launcher racks and a battalion could lay down an immense rapid-fire barrage.

The rocket was 1.8 m/5.9 ft long and 130 mm/5.1 in diameter and weighed 42 kg/92 lb complete with a 22 kg/48 lb explosive warhead.

Kesselring, Albert (1885–1960)
German **field marshal** in **World War II**, commander of the **Luftwaffe** (air

force) 1939–40, during the invasions of Poland and the Low Countries and during the early stages of the Battle of Britain. He later served under **Rommel** in North Africa, took command in Italy in 1943, and was commander-in-chief on the western front in March 1945. His death sentence for war crimes at the 1947 Nüremberg trials was commuted to life imprisonment, but he was released 1952.

Kharkov, Battle of

In **World War II**, series of battles 1941–43 between Soviet and German forces over possession of Kharkov, the fourth most important city in the USSR, about 480 km/300 mi east of Kiev. Kharkov was initially taken by the German 6th Army on 24 October 1941 with little resistance. In May 1942 a Soviet 'Front' of 640,000 troops and 1,200 tanks set out to recapture the city, but two German armies attacked the flanks and cut off the spearhead, taking about 250,000 prisoners and destroying the remainder. It was retaken by Soviet troops after their breakout from **Stalingrad** in February 1943 but was then recaptured by the German Army Group South on 15 March. General Rodion Malinovsky finally liberated it in August 1943, following the failure of the German offensive against **Kursk**.

Khe Sanh

In the **Vietnam War**, US Marine outpost near the Laotian border and just south of the demilitarized zone between North and South Vietnam. Garrisoned by 4,000 Marines, it was attacked unsuccessfully by 20,000 North Vietnamese (NVA) troops between 21 January and 7 April 1968.

The base was entirely surrounded and besieged, helicopters being the only means of supply. In order to reassure the South Vietnamese of US commitment, especially after NVA victories in the **Tet Offensive**, Gen William C Westmoreland, the US commander, decided to fight for Khe Sanh and provided heavy air support. Even so, a nearby smaller base held by Special Forces was overwhelmed by the NVA using light tanks, and the heavy artillery bombardment of Khe Sanh drove the defenders underground. A succession of NVA attacks was repulsed, and finally, on 1 April, a US attack to relieve the besieged base began. The siege was lifted on 7 April.

Kitchener, Horatio (Herbert), 1st Earl Kitchener of Khartoum (1850–1916)

Irish-born British soldier and administrator. He defeated the Sudanese at the

Battle of **Omdurman** in 1898 and reoccupied Khartoum. In South Africa, he was commander-in-chief between 1900 and 1902 during the **South African** or Boer War, where he conducted war by a scorched-earth policy and created the earliest concentration camps for civilians. He commanded the forces in India 1902–09. Appointed war minister on the outbreak of **World War I**, he was successful in his campaign of calling for voluntary recruitment. He planned for an entrenched three-year war, for which he began raising new armies. Kitchener drowned when his ship struck a German **mine** on the way to Russia.

> ❝ I don't mind your being killed, but I object to your being taken prisoner. ❞
>
> **Earl Kitchener of Khartoum**, to the Prince of Wales (later Edward VIII) when the prince asked to go to the Front in World War I, quoted in Viscount Esher's *Journal*, 18 December 1914.

knight

Medieval man-at-arms, who usually fought on horseback wearing **armour**. Norman knights defeated the English shield-wall at the battle of **Hastings** in 1066. French knights fell to hails of English arrows at the battles of **Crécy** and **Agincourt** during the **Hundred Years' War**. Knights fought in the **Crusades**, during which military-religious orders such as the Knights Templar came into being, but their effectiveness as heavy cavalry was ended by the introduction of **firearms** from the 1400s.

The oldest order of knighthood in the UK is the Order of the Garter, founded about 1347.

Kohima, Battle of

One of the most savage battles of **World War II** in April–May 1944 as the Allied garrison at Kohima, a town in Manipur province, northeast India, repulsed a wave of Japanese attacks, suffering severe casualties. It was attacked by the Japanese 31st Division on 5 April 1944, who isolated the British force at **Imphal**, about 95 km/60 mi to the north. The British 14th Division broke through and relieved the Kohima force on 14 April, but were unable to break through the Japanese positions blocking the route to Imphal until the end of May, by which time the Japanese were weakened by disease and starvation and began to retreat.

Korean War
War from 1950 to 1953 between North Korea (supported by China) and South Korea, aided by the United Nations (the troops were mainly US). North Korean forces invaded South Korea on 25 June 1950, and the Security Council of the United Nations, owing to a walkout by the USSR, voted to oppose them. The North Koreans held most of the South when US reinforcements arrived in September 1950 and forced their way through to the North Korean border with China. The Chinese retaliated, pushing them back to the original boundary by October 1950. Truce negotiations began in 1951, although the war did not officially end until 1953.

- By September 1950 the North Koreans had overrun most of the South, with the UN forces holding a small area, the Pusan perimeter, in the southeast.
- The course of the war changed after the surprise landing of US troops later the same month at **Inchon**, on South Korea's northwest coast.
- These troops, led by Gen Douglas **MacArthur**, fought their way through North Korea to the Chinese border in little over a month.
- On 25 October 1950 Chinese troops attacked across the Yalu River, driving the UN forces below the 38th parallel.

Kosovo, Battles of
Turkish victories over the combined forces of Serbia, Bosnia, and Albania in 1389 and 50 years later over a combined Hungarian and Wallachian army in 1448. The Plain of Kosovo forms part of Serbia and has always been a focal point for any military operations in the region.

The Serbian ruler was killed during the battle of 1389 and the Serbian Empire came under Turkish rule thereafter. The Serbians had their revenge during the Balkan War of 1912 when they soundly defeated the Turks in several small engagements in this area, and in 1915 it was the scene of the last stand of the Serbian Army before they were driven from the country.

Kosovo Liberation Army
KLA, also known as *Ushtria Çlirimtare e Kosovës (UCK)* an Albanian, paramilitary force operating in the predominantly ethnic Albanian province of Kosovo, Yugoslavia, and fighting for the independence of Kosovo. The

KLA emerged as an organized movement in 1996, and by 1998 found itself in command of an uprising, which quickly spread across parts of the province. Labelled a terrorist organization by the Serb authorities (and Russia), the KLA was attacked by the Serbs and by April 1999 – a month into a **NATO** offensive against Yugoslavia – the organization had been decimated, although thousands of new Kosovar recruits from European countries began to arrive. The KLA coordinated its operations with NATO air forces in its bombing campaign against Yugoslav military targets during 1999.

See also: *Yugoslav war.*

Kursk, Battle of

In **World War II**, unsuccessful German offensive against a Soviet salient in July 1943. Kursk was the greatest **tank** battle in history and proved to be a turning point in the Eastern Front campaign. With nearly 6,000 tanks and 2 million troops involved, the battle was hard fought, reaching its climax with the pitched battle on 12 July between 700 German and 850 Soviet tanks.

- The Soviets were forewarned and prepared for the assault with 20,000 guns, 3,300 tanks, 2,560 aircraft, and 1,337,000 troops.
- The Germans massed 10,000 guns, 2,380 tanks, 2,500 aircraft, and 900,000 troops.
- The Allied landing in Sicily on 10 July led **Hitler** to demand the withdrawal of troops from the USSR to reinforce Italy.
- Hitler terminated the Kursk battle on 17 July and German forces were left to extricate themselves as best they could.

Ladysmith

Town in province of KwaZulu-Natal (formerly Natal), South Africa. Founded in 1851, it was besieged for 118 days during the **South African War** or Boer War in 1899. During the **siege** 3,200 British soldiers were killed or died of disease. The Boers under Gen Joubert agreed to the evacuation of the women and children to Ntombi Camp, but British attempts to relieve the town by a column under Gen Sir Redvers Buller repeatedly failed. A sortie by 600 troops on 8 December captured a Boer gun, and the Boers made a determined attack on the town 6 January, but were repulsed. Eventually, at his 12th attempt, Buller succeeded in reaching the town and ended the siege on 28 February 1900.

landing craft

Specialized ships developed in **World War II** for the transport and delivery of troops and equipment in a sea-borne assault.

Landing ships were large craft capable of long sea voyages; they carried tanks, vehicles, and landing craft.

Landing craft were small flat-bottomed boats suitable only for use close inshore and were used to ferry troops and light equipment from landing ships and transports to the beach. They were used during the **D-Day** landings of 1944.

land mine

Explosive **mine** designed for use on land, often in large numbers in **minefields**, as a defence against attacking enemy vehicles and troops. Landmines are simple to make but very difficult to clear once an area has been mined, and they continue to pose a threat to civilians years after a conflict has ended. An estimated 100–200 million live mines are in the ground. Most of these are anti-personnel mines (APMs), designed to maim not kill. In 1997 more than 100 countries, including the UK, signed a draft treaty to ban anti-personnel mines immediately. The USA refused to sign the treaty.

- The country with the greatest number of uncleared mines is Afghanistan, with an estimated 9–10 million.
- There are 3 million land mines remaining in Bosnia.
- Cambodia has the highest number of amputees as a result of mine injuries, a proportion of one in every 236 people.

Lawrence, T(homas) E(dward) (1888–1935)
Known as 'Lawrence of Arabia', British soldier and scholar, famous for his role in the Arab revolt against the Turks during **World War I**. An intelligence officer in Cairo, Egypt, in 1916 he attached himself to Emir Faisal. He became a **guerrilla** leader of genius, combining raids on Turkish communications with the organization of a joint Arab revolt, later described in his book *The Seven Pillars of Wisdom* (1926). In 1918 he led the Arab army into Damascus.

Disappointed by the Paris Peace Conference's failure to establish Arab independence, he joined the Royal Air Force in 1922 as an aircraftman under the name Ross, transferring to the tank corps under the name T E Shaw in 1923 when his identity became known. In 1935 he was killed in a motorcycle accident.

> It's the most amateurish, Buffalo-Billy sort of performance, and the only people who do it well are the Bedouin.
>
> **T E Lawrence**, letter 1917, describing an attack on a Turkish train.

Lee, Robert E(dward) (1807–1870)
US military strategist and Confederate general in the American Civil War (*see* **Civil War, American**). Lee was born in Virginia. He graduated from West Point, was commissioned in 1829, and served in the Mexican War from 1846 to 1948. In 1859 he suppressed John Brown's raid on Harper's Ferry. On the outbreak of the Civil War in 1861 he joined the army of the Confederacy of the Southern States. As military adviser to Jefferson Davis, president of the Confederacy, and as commander of the Army of Northern Virginia, he made several raids into Northern territory, but was defeated at **Gettysburg** and he surrendered to Gen **Grant** on 9 April 1865 at Appomattox Court House.

- In 1862 Lee won the Seven Days' Battle defending Richmond, Virginia, the Confederate capital, against Gen George McClellan's Union forces.
- In 1863 Lee won victories at Fredericksburg and Chancellorsville.
- In 1864 he won again at Cold Harbor, but was besieged in Petersburg from June 1864 to April 1865.

Lee, Robert General Robert E. Lee bids farewell to his troops.

Lee-Enfield rifle
British service rifle. A .303-in calibre weapon, it used a bolt-action breech and a 10-round box magazine. It was first used 1893 and went through several minor modifications, but remained the standard British army rifle until the 1950s.

legion
Main unit of the Roman army for most of its history, originally a citizen force. At the time of the Second **Punic War** (218–201 BC), the legion included 300 **cavalry**, 1,200 light **infantry**, and 4,200 heavy infantry. All were equipped with a **helmet**, body **armou**r, oval **shield**, and *gladius* (a short **sword**). Some carried two *pila* (javelins) per man, while others were armed with long spears. The Roman army became a professional force in the late 2nd–early 1st centuries BC. The ancient property qualification was gradually abandoned and all legionaries were armed alike with long shield, helmet, cuirass, *gladius*, and *pilum*. Facts about the legion in the late imperial period are uncertain. During the 3rd century BC most were reduced in size, probably to about 1,000 strong.

- At its peak, the legion was organized into ten cohorts of 480 men; each divided into six centuries of 80, each century commanded by a centurion.
- A legate commanded legions.
- Soldiers served for 25 years and received a bounty or plot of land on discharge.

- There were many specialists such as engineers, artillerymen, architects, and artisans.

Leipzig

City in west Saxony, Germany, 145 km/90 mi southwest of Berlin, the scene of fighting in the **Thirty Years' War** (1618–48), the **Seven Years' War** (1756–63), and during the **Napoleonic Wars** in the period 1812–13. The Battle of the Nations, which saw the defeat of **Napoleon**, took place outside Leipzig in 1813.

Leningrad, Siege of

In **World War II**, German siege of the Soviet city of Leningrad (now St Petersburg, Russia) lasting from 1 September 1941 to 27 January 1944. The Germans reached Leningrad during the initial stages of their invasion of the USSR, Operation **Barbarossa**, and were prepared to take it, but **Hitler** ordered them to besiege the city in order to achieve a bloodless occupation. All land communication was cut off and the city subjected to air and artillery bombardment. Before the year was out, starvation was causing 300 deaths a day, although this was partially eased when Lake Ladoga froze, enabling a truck route to be established that brought food in over the ice. Some 1 million inhabitants of the city are reckoned to have died during the 900 days of the siege, either from disease, starvation, or enemy action.

Lexington and Concord, Battle of

First battle of the **American Revolution** on 19 April 1775 at Lexington, Massachusetts, (now USA). The first shots were fired when British general Thomas Gage sent 800 troops to seize stores at Concord and arrest rebel leaders John Hancock and Samuel Adams. About 50 local militia ('minutemen') attacked the British. Eight minutemen were killed and the remainder retired. The British party turned back for Concord and was later ambushed; it was only saved by reinforcements sent out from Concord. The total losses in the two actions were 73 British killed and 174 wounded, 49 Americans killed and 39 wounded. Although a somewhat inconclusive action in itself, it sparked wider rebellion and so precipitated the revolution.

Leyden, Siege of

During the Netherlands War of Independence, **siege** of the Dutch city of Leyden (now Leiden) from 26 May to 3 October 1574 by Walloon and German troops. The Dutch force was little more than the town guard and a few mercenaries, while the besiegers numbered about 8,000 and had ample **artillery**. To save the city, the Prince of Orange ordered the sea gates to be opened, bursting the dykes and flooding large areas surrounding the city. A Dutch fleet broke through the blockade at the end of September and the townspeople were then able to capture the remaining German **redoubt**, and the siege was broken on 3 October.

Leyte Gulf, Battle of

In **World War II**, US naval victory over Japan from 17 to 25 October 1944, to the east of the Philippines. The biggest naval battle in history, involving 216 US warships, 2 Australian vessels, and 64 Japanese warships, it resulted in the destruction of the Japanese navy.

There were a number of separate actions as elements of these forces met and diverged, but the overall result was total victory for the Allies. While the naval battle was raging, Gen Krueger landed the 6th US Army on Leyte Island, which was secured on 25 December 1944.

- The Japanese lost some 500 aircraft, 3 battleships, 4 carriers, 10 cruisers, 11 destroyers, and 1 submarine.
- Of the surviving Japanese warships, few if any escaped damage.
- The Americans lost 200 aircraft 2 destroyers, 1 destroyer escort, and 3 light carriers.

Liaoyang, Battle of

Inconclusive clash between Japanese and Russian forces during the **Russo-Japanese War** from 25 August to 4 September 1904, in Manchuria, about 80 km/50 mi southwest of Mukden (now Shenyang). The battle foreshadowed the tactics of **World War I**, with the use of massed armies, modern **rifles** and **machine guns**, barbed wire, and entrenchments.

The Russian strength was about 145,000 troops; the Japanese force, under Marshal Iwao Oyama, was about 120,000 strong between three armies, though the Russians believed it to be much stronger. The Russians retired in good order with their reserves uncommitted and lost only 16,000 troops to the Japanese losses of 25,000. Had their commander Kuropatkin kept his nerve, and had he known the actual Japanese strength, he could probably have defeated them.

Light Brigade, Charge of the
See **Charge of the Light Brigade.**

Little Bighorn, Battle of the

Battle in Montana, USA, one of the last battles of the Indian Wars. Lt-Col George Custer suffered a crushing defeat in the hands of the Sioux and Cheyenne American Indians on 25 June 1876, under chiefs Crazy Horse and Sitting Bull. The battle is also known as Custer's Last Stand.

Custer had been sent with a detachment of the US Seventh Cavalry to quell the Sioux uprising that had broken out in the Black Hills of South Dakota. Following the discovery of gold, white miners had been allowed to encroach on this area, which was sacred to the Sioux and had been granted to them in perpetuity by the US government as a homeland in 1868. Custer ignored scouting reports of an overwhelming Indian force and led a column of 265 soldiers into a ravine where thousands of Indian warriors lay in wait. In the battle, which lasted for just one hour, Custer and every one of his command were killed.

Devastating US reprisals against the Indians followed, driving them from the area.

Long Range Desert Group

Highly mobile British penetration force formed in July 1940 during **World War II** to carry out reconnaissance and raids deep in the desert of North Africa. After the successful conclusion of the North African campaign in 1942, the group was redeployed to carry out operations in Greece, Italy, and Yugoslavia. It was disbanded in August 1945. At its full strength in March 1942, it had 25 officers, 324 soldiers, and 110 vehicles.

Loos, Battle of

In **World War I**, unsuccessful French-British offensive of September 1915 to recover the mining districts around the towns of Loos and Lens from the Germans. This was the first action in which the British used gas (see **gas warfare**), but the wind shifted and blew the gas back over British lines. British forces were largely composed of fresh volunteers from the **Kitchener** armies. A German counterattack drove the British back to their start line, although a second attack further south gained them about 3 km/ 2 mi. Some 500,000 Allied troops were deployed in the battle and the British lost 60,000 casualties including three generals. French casualties are believed to have been even higher.

Lucknow, Siege of

During the Indian Mutiny, **siege** of the British residency (governor general's headquarters) in Lucknow, Uttar Pradesh, between 2 July and 16 November

1857. Over 500 British troops and civilians with 700 loyal Indian troops were besieged in the residency building for four months until a relief column finally broke through the mutineers' blockade. The city was not recaptured until March 1858.

> **BREAKING THE SIEGE**
>
> - A column led by Gen Henry Havelock managed to fight its way through on 25 September.
> - But the combined force was not strong enough to fight its way out again.
> - After 53 days, 5,000 more troops, led by Gen Colin Campbell, appeared outside the city.
> - Thomas Kavanagh, a civil servant, slipped through the rebel lines in disguise to guide the relief force. He became the first civilian to be awarded the Victoria Cross.
> - Campbell rescued the besieged force and civilians, and then left the city to the rebels.

Ludendorff, Erich von (1865–1937)

German general, Chief of Staff to Hindenburg in **World War I**, and responsible for the eastern-front victory at **Tannenberg** in 1914. After Hindenburg's appointment as chief of general staff and Ludendorff's as quartermaster-general in 1916, the two were largely responsible for the conduct of the war from then on. After the war he propagated the 'stab in the back' myth, according to which the army had been betrayed by the politicians in 1918. He took part in the Nazi rising in Munich in 1923 and sat in the Reichstag (parliament) as a right-wing Nationalist.

> ❝ The Army had been fought to a standstill and was utterly worn out. ❞
>
> **Erich von Ludendorff** on the Battle of the Somme.

Luftwaffe

German air force used both in **World War I** and (as reorganized by the Nazi leader Hermann Goering in 1933) in **World War II**. Germany was not

supposed to have an air force under the terms of the Treaty of Versailles 1918, so the Luftwaffe was covertly trained and organized using Lufthansa, the national airline, as a cover; its existence was officially announced on 1 April 1935. It was one of the vital components of the **Blitzkrieg** tactics. Although some officers advocated strategic long-range bombing, they were ignored, and except for maritime reconnaissance, the Luftwaffe never operated any long-range aircraft. The Luftwaffe was also responsible for Germany's anti-aircraft defences, operating both guns and aircraft.
See also: *Britain, Battle of.*

Lusitania
Ocean liner sunk in the Atlantic by a German **submarine** on 7 May 1915 with the loss of 1,200 lives, including some US citizens. Its destruction helped to bring the USA into **World War I**.

Lützen, Battle of
In the **Thirty Years' War**, Swedish victory on 16 November 1632 over an Imperial army under Albrecht von **Wallenstein**, 45 km/28 mi west of Leipzig, Germany. King Gustavus Adolphus who was killed during the battle led the Swedish army, of about 19,000 troops.

Lützen, Battle of
In the **Napoleonic Wars**, disastrous surprise attack in Germany on **Napoleon** Bonaparte, by a joint Prussian and Russian army under Count Wittgenstein on 2 May 1813.

Napoleon was moving toward the Elbe with about 200,000 troops and directed his advanced guard to Lützen. Wittgenstein decided to attack with a small detachment, whilst directing the major part of his army against Napoleon's right and rear. Napoleon, hearing Wittgenstein's **artillery**, realised what was intended. He withdrew a reserve, leaving the rest to hold off the Allies. When both sides had fought themselves to a standstill, he sent in a grand battery of 100 guns to blow a hole in the enemy line with case shot, through which he then threw his reserve force. The Prussians and Russians lost about 20,000 troops, the French about half that number.

MacArthur, Douglas (1880–1964)

US general in **World War II**, commander of US forces in the Far east and, from March 1942, of Allied forces in the southwest Pacific. He defended the Philippines against the Japanese in 1941-42 and was responsible for the re-conquest of New Guinea and the Philippines from 1944 to 1945. He was appointed general of the army in 1944. After the surrender of Japan he commanded the Allied occupation forces there. He was commander of UN forces in the **Korean War**, where his threats to bomb China aroused fears of a new world war. He was removed from command but received a hero's welcome on his return to the USA.

❝ It is fatal to enter any war without the will to win it. ❞

Gen Douglas MacArthur, speaking at the US Republican Convention, 1952.

machine gun

Rapid-firing automatic gun. The Maxim gun, named after its inventor, US-born British engineer H S Maxim (1840–1916), of 1884 was recoil-operated, but some later types have been gas-operated (**Bren**) or recoil assisted by gas (some versions of the Browning).

The forerunner of the modern machine gun was the **Gatling** (named after its US inventor R J Gatling, 1818–1903), perfected in the USA in 1860 and used in the **Civil War**. It had a number of barrels arranged about a central axis, and the breech containing the reloading, ejection, and firing

machine gun *Gunner operating a Maxim machine gun.*

Madrid, Siege of

In the Spanish **Civil War**, Nationalist siege of Republican forces in Madrid, from November 1936 to March 1939. The city was a stronghold of Republican opposition to the Nationalist leader, Gen Francisco Franco who, after a two-year siege, launched a sudden offensive against Madrid on 26 March 1939. Resistance collapsed and the city was occupied on 31 March, ending the Civil War.

See also: *fifth column*

The Nationalist commander Gen Emilio Mola famously boasted that he had four columns of troops outside the city and a 'fifth column' of sympathizers inside.

Mafeking, Siege of

Boer siege, during the **South African War**, of the British-held town (now Mafikeng) from 12 October 1899 to 17 May 1900. A 10,000-strong Boer force besieged the British garrison, of about 750 soldiers under Col Robert Baden-Powell, 1,700 townspeople, and about 7,000 Africans. The only serious attack the Boers attempted failed.

The announcement of Mafeking's relief led to wild scenes of celebration in Britain, and even led to the coining of a new verb – 'to maffick', meaning to celebrate intemperately.

Inside the town, Baden-Powell kept life proceeding more or less as normal; there were tea parties, concerts, and polo matches. Eventually a British column led by Col Herbert Plumer and Col Mahon arrived, and relieved the town on 17 May 1900.

Magdeburg, Sack of

During the **Thirty Years' War**, Imperial victory over the Swedes on 20 May 1631 at Magdeburg, Germany. The Swedes had captured Magdeburg 1629, holding the city with a small garrison until an Imperial force under Count Tilly arrived early in 1631. He besieged it for five months, finally storming the city in May 1631 to forestall the arrival of Swedish reinforcements.

Tilly's troops then sacked the city and massacred most of the population. Magdeburg itself was set on fire, leaving only the cathedral standing.

Maginot Line

French **fortification** system along the German frontier from Switzerland to Luxembourg built 1929–36 under the direction of the war minister, André Maginot. It consisted of semi-underground forts joined by underground passages, and was protected by antitank defences; lighter fortifications continued the line to the sea. In 1940 during **World War II** German forces pierced the Belgian frontier line and outflanked the Maginot Line.

Malplaquet, Battle of

During the War of the Spanish Succession, victory of the British, Dutch, and Austrian forces over the French forces on 11 September 1709 at Malplaquet, in Nord *département*, France. The Imperial army was under the command of the Duke of **Marlborough** and Prince **Eugène of Savoy** while the French were under Marshal Claude de Villars. No other battle during this war approached Malplaquet for ferocity and losses sustained by both sides – the joint Imperial force lost over 20,000 troops and the French 12,000, both having begun with about 90,000. Losses on the Imperial side were so great and the troops so fatigued, that they were unable to pursue the French as they retreated.

Manhattan Project

Code name for the development of the **atom bomb** in the USA during **World War II,** to which the physicists Enrico Fermi and J Robert Oppenheimer contributed.

Manzikert, Battle of

Decisive battle at Manzikert in eastern Turkey, on 19 August 1071, between the Seljuk Turks under Sultan Alp-Arslan and the Byzantines led by Emperor Romanus IV (1067–71). It resulted in the destruction of the Byzantine regular army and the loss of most of Anatolia to the Turks.

- The Byzantines formed a typical two-line formation, then advanced towards the Seljuk camp.
- Their opponents retired in the face of their approach.
- At dusk, Romanus began to withdraw towards his own camp. The order caused confusion in the Byzantine ranks.

- The Turks seized the opportunity to attack. Every man in the Byzantine front line was killed or captured, including the emperor who was taken prisoner.

Mao Zedong (Mao Tse-tung) (1893–1976)

Leader of the Chinese Communist Party (CCP) from 1935 to 1976. During the Chinese civil war between Communists and Nationalists, he led the **Red Army**, together with Zhu De, on the Long March north to Yan'an, in Shaanxi in 1934–35. He organized the war of liberation against the Japanese occupying forces, setting up an alliance with the nationalist Kuomintang in 1936. After 1945 and the defeat of Japan, the CCP Red Army successfully employed mobile, rural-based **guerrilla** tactics against the Kuomintang. Mao helped established a People's Republic and communist rule in China. He was state president until 1959, and headed the CCP until his death. His influence diminished with the failure of his 1958–60 'Great Leap Forward', but he emerged dominant again during the 1966–69 Cultural Revolution, which he launched to purge the party of 'revisionism'.

> ❛Communism is not love. Communism is a hammer, which we use to crush the enemy.❜
>
> **Mao Zedong**, quoted in *Time*, 18 December 1950.

Marathon, Battle of

One of the most famous battles of antiquity, fought in September 490 BC on the Plain of Marathon, northeast of Athens, at the start of the **Persian Wars**. The Athenians and their allies resoundingly defeated the Persian king Darius's invasion force. The Greeks were a combined force of about 10,000 Athenians under Miltiades supplemented by Plataeans. The Persian force numbered perhaps 25,000 in all, including cavalry. About 6,400 Persians were allegedly killed for the loss of only 192 Athenians. The victory was an enormous boost to Greek morale.

- The battle has been immortalized by the marathon race named in memory of the runner, Pheidippides.
- He reputedly ran to Sparta from Athens to appeal for aid before the battle. He covered the distance of 200 km/125 mi in a day but the Spartans failed to provide any assistance.

- A more recent legend, that he ran from Athens to Marathon (about 40 km/25 mi) to fight, then ran back with the news of the victory before dropping dead, is considered spurious by scholars.

Marengo, Battle of

During the **Napoleonic Wars**, defeat of the Austrians on 14 June 1800 by the French army under **Napoleon** Bonaparte, near the village of Marengo in Piedmont, Italy. It was one of Napoleon's greatest victories, and resulted in the Austrians ceding northern Italy to France.

Napoleon assembled an army of about 40,000 troops in Switzerland and marched them secretly across the St Bernard Pass into Italy. Eventually the Austrians found them, and the 28,000 French, outnumbered by an Austrian force of about 32,000, were falling back when Napoleon arrived. He organized the retreat in good order, until reinforcements arrived. French cavalry attacks and artillery then broke the Austrians who fled the field, losing about 9,000 troops. French losses were about 4,000.

marines

Fighting force that operates both on land and at sea.

The US Marine Corps (1775) is constituted as an arm of the US Navy. It is made up of infantry and air support units trained and equipped for amphibious landings under fire.

The British Corps of Royal Marines (1664) is primarily a military force, also trained for fighting at sea, and providing **commando** units, landing craft, crews, and **frogmen**.

Marlborough, John Churchill, 1st Duke of Marlborough (1650–1722)

English soldier, created a duke in 1702 by Queen Anne. He was granted the Blenheim mansion (Blenheim Palace) in Oxfordshire

Marlborough, John Churchill *A portrait of the Duke of Marlborough in ceremonial battledress.*

in recognition of his services, which included defeating the French army outside Vienna in the Battle of **Blenheim** in 1704, during the **War of the Spanish Succession**.

He achieved further victories in Belgium at the battles of **Ramillies** (1706) and Oudenaarde (1708) and in France at **Malplaquet** in 1709. However, the return of the Tories to power and his wife's quarrel with the queen led to his dismissal in 1711, and his flight to Holland to avoid charges of corruption. He returned to England in 1714.

> ❝ I have not time to say more, but beg you will give my duty to the Queen, and let her know her army has a glorious victory. ❞
>
> The **Duke of Marlborough**, in a letter to his wife, referring to the *Battle of Blenheim*, 1704.

Marne, Battles of the

In **World War I**, two unsuccessful German offensives in northern France. In the first battle, between 6 and 9 September 1914, German advance was halted by French and British troops under the overall command of the French general Joseph Joffre. Although tactically inconclusive, the first battle of the Marne was a strategic victory for the Allies. In the second battle, from 15 July and 4 August 1918, the German advance was defeated by British, French, and US troops under the French general Henri Pétain, and German morale crumbled. They were halted and turned back. The Allied counterattack, beginning 18 July, is sometimes referred to as the Third Battle of the Marne.

marshal

In the French army the highest officers bear the designation of *maréchal de France*/marshal of France. It corresponds to admiral of the fleet in the navy, and field marshal in the army. Leading generals of **Napoleon** Bonaparte were appointed marshals, such as Michel **Ney**. Marshal is also the highest rank in the British Royal Air Force.

Mauser rifle

Service **rifle** of the German Army in **World War I**, adopted in 1898. A 7.92 mm/3 in calibre, bolt-action rifle, it had a five-shot magazine.

The original design, the Gewehr '98, was a full-length rifle. It was supplemented by a short-barrelled version known as the Karabiner '98, which gradually replaced the longer weapon and became the standard rifle for all arms of the services.

medals and decorations
Coinlike metal pieces, struck or cast to commemorate historic events, to mark distinguished service whether civil or military (in the latter case in connection with a particular battle, or for individual feats of courage, or for service over the period of a campaign), or as a badge of membership of an order of knighthood, society, or other special group. Notable awards for bravery in battle include the French Croix de Guerre, the German Iron Cross, the British Victoria Cross, the US Medal of Honor and, (the oldest US decoration, established 1782) the Purple Heart.

mercenary
Soldier hired by the army of another country or by a private army. Mercenary military service originated in the 14th century, when cash payment on a regular basis was the only means of guaranteeing soldiers' loyalty. Most famous of the mercenary armies of that time was the Great Company, which was in effect a glorified protection racket, comprising some 10,000 knights of all nationalities and employing condottieri, or contractors, to serve the highest bidder. In the 18th century, Swiss cantons and some German states regularly provided the French with troops for mercenary service. Britain employed 20,000 German mercenaries to make up its numbers during the **Seven Years' War** (1756–63) and used Hessian forces during the **American Revolution** (1775–83). In the 20th century mercenaries were common in wars and **guerrilla** activity in Asia, Africa, and Latin America.

Meuse, Battles of the
In **World War I**, battles between French and German forces in August 1914 on the line of the River Meuse in northern France. The French stemmed at least part of the German invasion but did not fully exploit their advantage.

The Germans were driven back over the Meuse, a considerable victory for the French. However, they failed to exploit it, as Marshal Joffre ordered a general retreat and the French commander in the field promptly obeyed, pulling back rapidly and abandoning large areas which could have been used to delay the subsequent German advance.

midget submarine

Small one- or two-person submersible capable of carrying one or two torpedoes. Midget submarines were used by the UK (see **X-craft**), Germany, Japan, and Italy in **World War II** for entering restricted waters inaccessible to conventional **submarines**.
See also: *human torpedo.*

Midway, Battle of

In **World War II**, decisive US naval victory over Japan in June 1942 off Midway island, northwest of Hawaii. In May 1942 the Japanese planned to expand their conquests by landing troops in the Aleutian Islands and on Midway. The Aleutian force was to draw the US fleet north, allowing the Midway force a free hand. The US forces deciphered Japanese naval codes and were able to intercept the mission. Both launched aircraft and the Americans sank one Japanese carrier and so damaged another two that they were abandoned. The sole remaining Japanese carrier managed to launch a strike that sank the USS *Yorktown* on 7 June, but later in the day another US strike damaged it so badly that it had to be scuttled. With no aircraft carriers or aircraft left, the Japanese retreated. Midway was one of the most important battles of the war in the Pacific, putting an end to Japanese expansion.

> The wreck of the USS *Yorktown* was found 1998, 5 km/3 mi down on the floor of the Pacific Ocean.

Minden, Battle of

During the **Seven Years' War**, French defeat by a combined British and Hanoverian army on 1 August 1759 at Minden, 70 km/44 mi west of Hanover, Germany. To repel a French advance, the Hanoverian commander Prince Frederick of Brunswick launched a counterattack of six English and three Hanoverian battalions of **infantry** moving steadily forward in line. They were attacked by the French cavalry, whom they annihilated with close-range **musket** fire, and the battle was won. Due to mismanagement of the victorious **cavalry**, the

> More than half of the losses to the British-Hanoverian side, were in the six English battalions, the descendants of which still wear a rose in their caps on the anniversary of the battle.

French were able to withdraw in good order, but at a loss of over 7,000 casualties and 43 guns. British and Hanoverian losses were some 2,700.

mine

Explosive charge on land or sea, or in the atmosphere, designed to be detonated by contact, vibration (for example, from an enemy engine), magnetic influence, or a timing device. Countermeasures include metal detectors (useless for plastic types), specially equipped **helicopters**, and special ships known as **minesweepers**.

The term originally denoted a tunnel driven beneath an enemy position and packed with **explosives** which were then detonated to coincide with an attack, a tactic used in many places on the Western Front in **World War I**. Mines were first used at sea in the early 19th century, during the **Napoleonic Wars**. **Landmines** came into use during World War I to disable **tanks**.

minefield

Area where mines have been planted to trap an unwary enemy or to prevent them from crossing or gaining access to the area, at sea or on land.

minesweeper

Small naval vessel for locating and destroying **mines** at sea. A typical minesweeper weighs about 725 tonnes/713 tons, and is built of reinforced plastic (immune to magnetic and **acoustic** mines). Remote-controlled miniature **submarines** may be used to lay charges next to the mines and destroy them.

Minesweeping was originally carried out by dragging some form of cable device through the water, which would catch the cables attaching the mines to their 'sinkers', pulling the mines up, or break the cables, thus allowing the mines to float up. Once on the surface, the mines could be detonated safely by gunfire.

MIRV

Abbreviation for multiple independently targeted re-entry vehicle. A MIRV is the nuclear-warhead carrying part of a ballistic **missile** that splits off in Midori from the main body. Since each is individually steered and controlled, MIRVs can attack separate targets over a wide area.

The US, former Soviet Union, UK, and French nuclear missiles are all equipped with MIRVs.

missile

Rocket-propelled weapon, which may be nuclear-armed (see **nuclear warfare**). Modern missiles are often classified as surface-to-surface missiles (SSM), air-to-air missiles (AAM), surface-to-air missiles (SAM), or air-to-surface missiles (ASM). A **cruise missile** can be sea-launched from submarines or surface ships, or launched from the air or the ground. Strategic missiles are the large, long-range, intercontinental ballistic missiles (ICBMs, capable of reaching targets over 5,500 km/3,400 mi). Tactical missiles are the short-range weapons intended for use in limited warfare (with a range under 1,100 km/680 mi).

There are many missiles that are small enough to be carried by one person. The Stinger, for example, is an anti-aircraft missile fired by a single soldier from a shoulder-held tube. Most **fighter** aircraft are equipped with air-to-air missiles for combat, and also air-to-surface weapons. Ship-to-ship missiles such as the Exocet have proved effective against warships.

- Rocket-propelled weapons were first used by the Chinese in about AD 1100.
- The rocket missile was re-invented by William Congreve in England around 1805, and remained in use with various armies in the 19th century.
- The first wartime long-range missile was the jet-powered German **V1** (Flying Bomb) of **World War II**.
- The first rocket-propelled missile with a pre-set guidance system was the German **V2**, also in World War II.

Missolonghi, Battle of

During the Greek War of Independence, Turkish victory over the Greeks on 22 April 1826, at the town of Missolonghi northwest of Patrai, Greece.

The Turks had besieged the town in 1821, but after two months the Turks had withdrawn. They returned in 1825 and began another siege, but in spite of several months of pressure the town failed to capitulate. Eventually the Turks gained reinforcements from an Egyptian army under Ibrahim Pasha, but it was a further three months before they were able to take the town by storm. The capture of Missolonghi finally moved Britain, France, and Russia to come to the aid of the Greeks.

Molotov cocktail

Popular name for a petrol bomb, a homemade weapon consisting of a bottle filled with petrol, plugged with a rag as a wick, ignited, and thrown as a grenade. Resistance groups during **World War II** named them after the Soviet foreign minister Molotov.

Moltke, Helmuth Carl Bernhard, Count von Moltke (1800–1891)

Prussian general responsible for devising the strategy that brought Prussia swift victories in the wars with Denmark in 1863–64, Austria in 1866 (the 'Seven Weeks' War'), and France in 1870–71 (*see* **Franco-Prussian War**). His reforms included training an elite body of staff officers and new emphasis on military intelligence. His most far-reaching innovation was his use of railways for rapid mobilization.

Moltke's nephew Helmuth von Moltke followed his uncle as chief of the general staff of the Prussian army (1906–14), but was dismissed after the German defeat by France at the First Battle of the **Marne** in **World War I**.

> *Everlasting peace is a dream, and not even a pleasant one ... war is a necessary part of God's arrangement of the world.*
>
> **Count Helmuth von Moltke**, in a letter to Dr J K Bluntschli, 11 December 1880.

Mons, Battle of

In **World War I**, German victory over the **British Expeditionary Force**, in August 1914. A planned attack on the German armies invading Belgium fell apart when French troops did not arrive, leaving the British to extricate themselves as best they could. They were forced out of a prepared defensive position and gradually retreated southward in a series of leap-frogging rearguard actions until the Germans over-stretched their supply lines and slackened their pursuit. The British formed a fresh line and the retreat was over.

Montgomery, Bernard Law, 1st Viscount Montgomery of Alamein (1887–1976)

English field marshal and a leading commander during **World War II**. Montgomery commanded part of the **British Expeditionary Force** in France 1939–40 and took part in the evacuation from **Dunkirk**. In August 1942 he took command of the 8th Army, then barring the German advance on Cairo. The victory of El **Alamein** in October turned the tide in North Africa and it was followed by the expulsion of **Rommel** from Egypt. In February 1943 Montgomery's forces came under US general **Eisenhower's** command, and they took part in the conquest of Tunisia and Sicily and the invasion of Italy. Montgomery commanded the Allied armies during the opening phase of the invasion of France in June 1944. At his headquarters on Lüneburg Heath, he received the German surrender on 4 May 1945.

mortar

Method of projecting a **bomb** via a high trajectory at a target up to 6–7 km/3–4 mi away. A mortar bomb is stabilized in flight by means of tail fins. The high trajectory results in a high angle of attack and makes mortars more suitable than **artillery** for use in built-up areas or mountains, even though they are not as accurate. Artillery also differs in that it fires a projectile through a rifled barrel, thus creating greater muzzle velocity.

Mortars began to be developed when the trench lines came into use in World War I, so that missiles could be pitched into the enemy trenches.

Moscow, Battle for

In **World War II**, a failed German attack on Moscow, from October 1941 to January 1942. The Soviet capital was a prime objective of the German invasion plan, Operation **Barbarossa**.

The Germans advanced within 24 km/15 mi of the city centre but the bitter cold and a fuel shortage halted them on 5 December. By this time, Marshal Georgi **Zhukov** had been appointed commander-in-chief of the defence of Moscow, and he mounted a powerful counterattack on 6 December. By 15 January 1942, the Soviet forces had pushed the Germans back to a line about 160 km/100 mi west of Moscow. The failure to capture the city was a severe setback for the German strategy.

Mukden, Battle of

Japanese victory over the Russians, the last battle of the **Russo-Japanese War**, from February to March 1905, outside Mukden (now called Shenyang), capital city of Manchuria. After the Battle of **Liaoyang** in September 1904, the Russians fell back to a defensive line south of Mukden. Two Japanese armies began attacking in February, forcing the two ends of the Russian defensive line to curve backwards. Fearing the Japanese would encircle the city, the Russians began a general retreat, which deteriorated into a total collapse. Mukden was evacuated by 10 March.

- Russian casualties were 26,500 killed, about the same number wounded, and 40,000 taken prisoner.
- The Japanese lost 41,000 killed and wounded.
- The Russian defeat finally persuaded the tsar to accept US mediation in June 1905.

Mulberry Harbour
Prefabricated floating harbour, used on **D-Day** in **World War II**, to assist in the assault on the German-held French coast of Normandy. Two were built in the UK and floated across the English Channel.

multiple independently targeted re-entry vehicle
See **MIRV**.

musket
Hand-held **firearm** developed from the arquebus. Muskets fired a lead ball, and were loaded from the muzzle end of the long barrel. The early matchlocks, in which a smouldering cord set off the **gunpowder** charge, were followed by flintlocks and wheel locks, which had more sophisticated mechanical devices to set off a spark. Muskets remained in use from the 1500s to the 1800s, and in pitched battles were used by ranks of **infantry** troops to fire volleys at one another.

musket *Musketeer, infantryman armed with a musket.*

mutiny

Organized act of disobedience or defiance by two or more members of the armed services. In naval and military law, mutiny has always been regarded as one of the most serious of crimes, punishable in wartime by death.

Effective mutinies in history include the Indian Mutiny by Bengal troops against the British in 1857 and the mutiny of some Russian soldiers in **World War I** who left the eastern front for home and helped to bring about the Russian Revolution of 1917. Most combatants in World War I suffered mutinies about the same time, the most serious outbreak affecting the French Army when whole battalions refused to fight.

In the UK, as defined in the 1879 Army Discipline Act, the punishment in serious cases can be death. The last British soldier to be executed for mutiny was Private Jim Daly, in India, in 1920.

My Lai massacre

Killing of 109 civilians in My Lai, a village in South Vietnam, by US troops in March 1968. An investigation in 1969 produced enough evidence to charge 30 soldiers with **war crimes**, but the only soldier convicted was Lt William Calley, commander of the platoon. Sentenced to life imprisonment 1971, Calley was released less than five months later on parole. The trial revealed a US Army policy of punitive tactics against civilians, and the massacre and its aftermath contributed to domestic pressure for the USA to end its involvement in the **Vietnam War**.

napalm
Fuel used in **flame-throwers** and **incendiary bombs**. Produced from jellied petrol, it is a mixture of *na*phthenic and *palm*itic acids. Napalm causes extensive burns because it sticks to the skin even when aflame. It was widely used by the US Army during the **Vietnam War**, and by Serb forces in the civil war in Bosnia-Herzegovina.

Napoleon I (1769–1821)
Napoleon Bonaparte, Emperor of the French from 1804–14 and from March to June 1815. Born in Ajaccio, Corsica, he received a commission in the artillery in 1785. A general from 1796 in the **French Revolutionary Wars**, in 1799 he overthrew the ruling Directory and made himself dictator. From 1803 he conquered most of Europe (these are known as the **Napoleonic Wars**). In 1804 a plebiscite made him emperor. In 1796 Napoleon had married Josephine de Beauharnais, but in 1809, to assert his equality with the Habsburgs, he divorced her to marry the Austrian emperor's daughter, Marie Louise. After the **Peninsular War** and retreat from Moscow in 1812, he was forced to abdicate in 1814 and was banished to the island of Elba. In March 1815 he reassumed power but was defeated by British and Prussian forces at **Waterloo** and exiled to the island of St Helena, where he died. His body was brought back in 1840 to be interred in the Hôtel des Invalides, Paris.

- Napoleon first distinguished himself at the siege of Toulon 1793.
- Having suppressed a royalist uprising in Paris 1795, he was given command against the Austrians in Italy.
- His fleet was destroyed by Nelson at the Battle of Aboukir Bay 1798.
- He defeated the Austrians at Marengo 1800.
- Prevented by the British navy's victory at Trafalgar from invading Britain, Napoleon drove Austria out of the war by victories at Ulm and Austerlitz 1805, and Prussia by the victory at Jena 1806.
- In 1812, Napoleon marched on Russia and occupied Moscow, but was forced to retreat.

- He was defeated at Leipzig 1813. Despite his brilliant campaign on French soil, the Allies invaded Paris and compelled him to abdicate in April 1814.
- In March 1815 he escaped and took power for a hundred days, with the aid of Ney, but lost finally at Waterloo 1815.

> ❝ Every French soldier carries in his cartridge-pouch the baton of a marshal of France. ❞
>
> **Napoleon I**, quoted in *Blaze La Vie militaire sous l'empire*.

Napoleonic Wars

Series of European wars (1803–15) conducted by **Napoleon I** of France against an alliance of Britain, the German states, Spain, Portugal, and Russia, following the **French Revolutionary Wars**, and aiming for French conquest of Europe. At one time nearly all of Europe was under Napoleon's domination. He was finally defeated at **Waterloo** in 1815.

During the Napoleonic Wars, the annual cost of the British army was between 60% and 90% of total government income. About half of Napoleon's army was made up of foreign **mercenaries**, mainly Swiss and German.

Narvik

Seaport on Ofot Fjord, north Norway, the scene of a naval action in **World War II**. To secure the supply of iron ore mined in the region, Germany seized Narvik in April 1940. On 13 April 1940 a British flotilla forced its way into Narvik Fjord and sank four German **destroyers**. British, French, Polish, and Norwegian forces recaptured the port on 28 May but had to abandon it on 10 June to cope with the deteriorating Allied situation elsewhere in Europe. Narvik was destroyed during the war but was rebuilt.

Naseby, Battle of

Decisive battle of the English **Civil War** on 14 June 1645, when the Royalists ('Cavalier's), led by Prince Rupert, were defeated by the Parliamentarians ('Roundheads') under Oliver **Cromwell** and Gen Fairfax. It is named after the nearby village of Naseby, south of Leicester.

NASEBY · 131

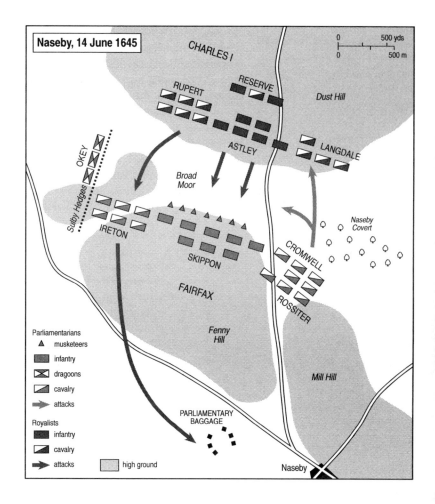

Both armies drew up in similar formation (*see* map). Prince Rupert's **cavalry** broke the Parliamentary right wing and then recklessly pursued them towards the village of Naseby. Cromwell's cavalry routed the force opposing them and then turned inward to take the Royalist **infantry** in the flank. King Charles I ordered his last reserves to charge, but the Earl of Carnwath, seeing this to be a futile move, turned his horse away and led his troops off the field; the Parliamentarians rallied, and completed the victory. Prince Rupert, returning from his chase, found the battle over.

NATO
Abbreviation for North Atlantic Treaty Organization, an alliance set up in 1949 to provide for the collective defence of western Europe and North America. Its military headquarters SHAPE (Supreme Headquarters Allied Powers, Europe) is in Belgium. During the **Cold War**, the NATO alliance was rivalled by the Warsaw Pact, a military grouping of Communist Eastern bloc states. When the Warsaw Pact was disbanded in 1991, several former members sought to join NATO. NATO forces acted as an intervention and peacekeeping force during the war in **Yugoslavia** 1999, bombing targets in Serbian territory, and sending ground troops to the Balkans region.

Navarino, Battle of
During the Greek war of liberation from Turkish rule, destruction on 20 October 1827 of a joint Turkish–Egyptian fleet by the combined fleets of the British, French, and Russians under Vice-Admiral Edward Codrington (1770–1851). The destruction of their fleet left the Turks highly vulnerable in Greece as they had no protection to their rear and no supply line, and this proved to be the decisive battle of the war. Navarino is the Italian and historic name of Pylos Bay, Greece, on the southwest coast of the Peloponnese.

> The battle began when a Turkish ship fired on a British gunboat. A general close-quarter battle broke out and within two hours the Turkish–Egyptian fleet had been almost totally destroyed.

navy
Fleet of ships, usually a nation's warships and the organization to maintain them. The world's most powerful navy is that of the USA. Other effective forces are those of Russia, France, and the UK. Such navies are equipped with **aircraft carriers**, **destroyers**, **frigates**, **submarines**, and smaller vessels, including supply ships and **assault ships**. Some navies maintain an air arm, with fixed wing aircraft and helicopters.

Nelson, Horatio, 1st Viscount Nelson (1758–1805)
English admiral, who was already a national hero before his death at the moment of victory at **Trafalgar** 1805. He joined the navy in 1770. During the **French Revolutionary Wars** he lost the sight in his right eye in 1794 and lost his right arm in 1797. He became a rear admiral and a national hero after the victory off **Cape St Vincent**, Portugal. In 1798 he tracked the French

fleet to **Aboukir Bay** in Egypt, and almost entirely destroyed it. In 1801 he won a decisive victory over Denmark at the Battle of **Copenhagen**, and in 1805, after two years of blockading Toulon, he defeated the Franco-Spanish fleet at the Battle of Trafalgar, near Gibraltar, capturing 20 of the enemy ships. Nelson himself was mortally wounded; he is buried in St Paul's Cathedral, London.

Nelson *Death of Admiral Nelson, Trafalgar, 1805.*

❝ England expects every man will do his duty. ❞

Nelson, before the Battle of Trafalgar, 1805.

New Orleans, Battle of

In the British–American **War of 1812**, battle between British and American forces December 1814–January 1815, at New Orleans. The war was already over by the time the battle was fought – peace had been signed on 24 December 1814 – but neither of the two forces in the area had received the news.

A garrison of about 6,000 troops under Gen Andrew Jackson held the city. A British fleet overpowered the American warships on the Mississippi River on 13 December and landed a force of about 6,000 British troops, who launched a determined attack on 1 January 1815. The assault failed, largely due to mismanagement and argument between the various commanders. Another attack on 8 January also failed; the British lost 1,500 troops, among them their commander Gen Pakenham.

Ney, Michel (1769–1815)

French soldier, Duke of Elchingen, Prince of Ney, **Marshal** of France under

Napoleon I. Ney commanded the rearguard of the French army during the retreat from Moscow, and for his personal courage at the Battle of Friedland in 1807 was called 'the bravest of the brave'. When Napoleon returned from exile in Elba, Ney was sent to arrest him, but instead deserted to the former emperor and fought at **Waterloo** in 1815. He was subsequently shot for treason.

The son of a cooper, he joined the army in 1788, and rose quickly within the ranks. He served throughout the **French Revolutionary Wars** and **Napoleonic Wars**.

Nile, Battle of the
Alternative name for the naval battle of **Aboukir Bay**.

Nimitz, Chester William (1885–1966)
US admiral, in **World War II** he was commander-in-chief of the US Pacific fleet. On 17 December 1941, following the Japanese attack on **Pearl Harbor**, the naval high command was overhauled and Nimitz took over command of the Pacific fleet. He directed the initial stages of the invasion of **Okinawa** in 1945. He reconquered the Solomon Islands in 1942–43, Gilbert Islands in 1943, the Mariana Islands and the Marshall Islands in 1944, and signed the Japanese surrender in 1945 as the US representative.

No Man's Land
During **World War I**, the space between the opposing trenches of the combatants. It is an old term for any piece of waste or unowned ground, used at least as early as the 14th Century.

Nuclear Non-Proliferation Treaty
Treaty signed in 1968 to limit the spread of nuclear weapons. Under the terms of the treaty, those signatories declared to be nuclear powers (China, France, Russia, the UK, and the USA) pledged to work towards nuclear **disarmament** and not to supply military nuclear technology to non-nuclear countries, while other signatories pledged not to develop or acquire their own nuclear weapons. The treaty was renewed and extended indefinitely in May 1995.

- Only 13 countries have not signed the treaty.
- In 1998 both India and Pakistan (non-signatories) became nuclear powers, after successful testing of weapons.

nuclear warfare

War involving the use of nuclear weapons. Nuclear-weapons research began in Britain in 1940, but was transferred to the USA after it entered **World War II**. J Robert Oppenheimer directed the research programme, known as the **Manhattan Project**. The original nuclear weapon, the **atom bomb**, relied on use of a chemical explosion to trigger a chain reaction. The first test explosion was at Alamogordo, New Mexico, on 16 July 1945; the first use in war was by the USA in World War II against Japan on 6 August 1945, over **Hiroshima** and three days later at **Nagasaki**. After the experience of World War II, the threat of nuclear war, the theory of **deterrence**, and the issue of **disarmament**, became key features of the **Cold War**.

NUCLEAR THREAT

- The worldwide total of nuclear weapons in 1990 was about 50,000.
- The number of countries possessing nuclear weapons stood officially at five – USA, USSR, UK, France, and China.

Okinawa

Group of islands, 520 km/323 mi from the Japanese mainland, forming part of the Japanese Ryukyu Islands in the west Pacific, the scene of a fierce battle during **World War II**. The main island, Okinawa, was captured by US forces between 1 April and 21 June 1945, with 47,000 US casualties (12,000 dead) and 60,000 Japanese (only a few hundred survived as prisoners). During the invasion over 150,000 Okinawans, mainly civilians, died; many were massacred by Japanese forces. The island was returned to Japan in 1972.

Omdurman, Battle of

Victory in September 1898 of British and Egyptian troops under **Kitchener** over Sudanese tribesmen (Dervishes) led by the Khalifa Abdullah el Taashi. The Khalifa was the successor to the Mahdi Mahomet Ahmed, who had fomented a revolt of Sudanese tribes against Egyptian rule, and become the unofficial ruler of southern Sudan. The British force sent to deal with him advanced slowly up the Nile.

> When the Battle of Omdurman ended it was found that the British rearguard troops had an average of two bullets each left in their pouches.

On 2 September the British were attacked by about 50,000 Dervishes, of whom an estimated 10,000 were killed. Kitchener ordered a march on to Omdurman but as the column moved off, a force of 20,000 Dervishes fell upon the rearguard. These troops stopped the Dervishes, killing most and scattering the remainder. The Khalifa escaped, to be pursued and later brought to battle and killed.

Opium Wars

Two wars, the First Opium War of 1839–42 and the Second Opium War of 1856–60, waged by Britain against China to enforce the opening of Chinese ports to trade in the narcotic drug opium. British military might proved too much for obsolete Chinese weaponry.

The First Opium War resulted in the cession of Hong Kong to Britain and the opening of five treaty ports. The Second Opium War followed with Britain and France in alliance against China, when there was further Chinese resistance to the opium trade. China was forced to give the European states greater trading privileges.

Orléans, Battle of

During the **Hundred Years' War**, English defeat by the French between October 1428 and May 1429. The English were rapidly conquering France and about 5,000 troops under the Earl of Salisbury attempted to take Orléans on 12 October 1428. The attempt failed, the Earl was killed, and the English laid **siege** to the city. They could not seal off the city completely, so there was no chance of starving out the French, and the siege lingered on until April 1429, when **Joan of Arc** arrived in Orléans. She took charge of the garrison and led them in a series of attacks on different English positions, taking them one after another until the remaining English raised the siege on 4 May 1429 and departed. The victory was a turning point in French fortunes in the war.

OSS

Abbreviation for Office of Strategic Services, a US secret intelligence agency during **World War II**. It was formed in 1942 to gather intelligence and carry out psychological and **guerrilla** warfare. OSS activities included the encouragement of resistance in **Axis**-occupied territories and sabotage. It was wound up when the war ended, and its activities were later continued by the Central Intelligence Agency (CIA), formed in 1947.

Pacific War
A war fought in 1879–83 by an alliance of Bolivia and Peru against Chile. Chile seized Antofagasta and the coast between the mouths of the rivers Loa and Paposo. This rendered Bolivia landlocked, and also annexed the south Peruvian coastline from Arica to the mouth of the Loa, including the nitrate fields of the Atacama Desert.

> **STILL UNRESOLVED**
>
> - Bolivia has since tried to regain Pacific access, either by a corridor across its former Antofagasta province or by a twin port with Arica at the end of the rail link from La Paz. Brazil supports the Bolivian claims, which would facilitate its own transcontinental traffic.

Pacific War
War fought in the Pacific theatre, both on land and at sea, between Japan and the Allies during **World War II**.
See also: *battles of Coral Sea, Midway, and Guadalcanal; Battle for Guam, Siege of Singapore; Sino-Japanese wars.*

Panipat, Battles of
Three decisive battles in the vicinity of this Indian town, about 120 km/75 mi north of Delhi:
- *27 April 1526* A smaller but highly trained force of Moguls under Babur, later the Mogul Emperor of India, defeated the Muslims under Ibrahim. The victory removed the Afghan dynasty from the Delhi throne and opened the way for the Mogul dynasty.
- *5 November 1556* An army of 100,000 Hindus, which had captured Delhi, was met by a Mogul army of about 20,000 troops and totally defeated, restoring Delhi to the Moguls and confirming Akbar as Emperor.

PARIS, SIEGE OF · 139

- *7 January 1761* During the Afghan–Mahratta wars, a Mahratta force of about 80,000 was defeated by an Afghan/Hindu army of similar strength.

Panzer

A German mechanized unit, created by Heinz **Guderain,** that was used in **World War II**. A Panzer (German 'armour') army was based on a core of tanks supported by infantry, artillery, and service troops in vehicles capable of accompanying tanks.

paratroops

Contraction of parachute troops, specially trained soldiers who are dropped by aircraft and parachute down into combat zones or to wherever they are needed. In war, they are often landed behind enemy lines and assigned to blow up bridges, destroy communications bases, or cut off supplies and reinforcements. They were used effectively by both the Allies and the Axis powers in **World War II** and have played a part in many military operations since then.

Paris, Siege of

Prussian siege of the city of Paris in 1870–71 during the Franco–Prussian War. After the defeat of the French armies in the field, Paris was the last vestige of French resistance. Count Helmuth von **Moltke** invested the city in September 1870 with 240,000 troops and upwards of 300 siege guns as well as field artillery. The bombardment, combined with shortage of food and a

Paris, Siege of *A poster calling on the citizens of Paris to take up arms against Prussian forces besieging the city.*

Passchendaele, Battle of

A successful but costly British operation in **World War I** to capture from the Germans the Passchendaele ridge in western Flanders from October to November 1917. British casualties numbered more than 300,000.

The ridge was an important strategic target of the British during the third Battle of **Ypres** (July–November 1971), as it gave them command of the Allied lines. It was retaken by the Germans in March 1918 and recovered again by the Belgians in October 1918.

> **YPRES END**
>
> The name Passchendaele is often erroneously applied to the whole of the third Battle of Ypres, but Passchendaele was in fact just the final part of it.

Patay, Battle of

A French victory by **Joan of Arc** over the English on 19 June 1429 at the village of Patay, 21 km/13 mi northwest of **Orléans**, during the **Hundred Years' War**.

The English sent an army, led by Sir John Talbot, to reinforce the siege of Orléans. They reached Patay but were discovered by some French scouts hunting a stag who informed Joan of Arc. The French fell on the English before they could gather themselves into a fighting formation. Talbot was taken prisoner and the English advance guard scattered, but the main body was able to make an orderly retreat back towards Paris.

Patriot missile

A ground-to-air, medium-range missile system used in air defence. It has high-altitude coverage, electronic jamming capability, and excellent mobility. US Patriot missiles were tested in battle against Scud missiles fired by the Iraqis in the 1991 Gulf War. They successfully intercepted 24 Scud missiles out of about 85 attempts.

> The Patriot missile was developed in the USA during the late 1970s. It entered service with NATO in West Germany in 1985.

Patton, George Smith (1885–1945)

US general in **World War II**, known as 'Old Blood and Guts'. In 1918, during **World War I**, he formed the first US tank force and led it in action. He was appointed to command the 2nd Armored Division in 1940 and became commanding general of the First Armored Corps in 1941. In 1942 he headed the Western Task Force that landed in Morocco, at Casablanca. After commanding the 7th Army in the invasion of Sicily, Patton led the 3rd Army across France and into Germany, and in 1945 took over the 15th Army.

> There's one thing you men can say when it's all over and you're home once more.
> You can thank God that twenty years from now when you're sitting by the fireside with your grandson on your knee, and he asks you what you did in the war, you won't have to shift him to the other knee, cough and say, 'I shovelled shit in Louisiana.'

George Patton, speech to US 5th Army prior to D-Day landings, 6 June 1944.

Pearl Harbor

US Pacific naval base on Oahu island, Hawaii, USA, the scene of a Japanese aerial attack on 7 December 1941, which brought the USA into **World War II**. The attack took place while Japanese envoys were holding so-called peace talks in Washington. More than 2,000 members of the US armed forces were killed, and a large part of the US Pacific fleet was destroyed or damaged.

The Japanese had been angered by US embargoes of oil and other war material and were sure that US entry into the war was inevitable. They hoped, by the attack, to force US concessions. Instead, it galvanized public opinion and raised anti-Japanese sentiment to fever pitch. War was declared shortly thereafter.

Peenemünde

A fishing village in Germany used from 1937 to develop the **V2** rockets deployed by the Germans in **World War II**. The RAF bombed it in August

1943, causing considerable damage and killing 750 staff including some crucial scientists, which set the programme back several months.

Peloponnesian War

A war fought between Athens and Sparta and their respective allies in 431–404 BC. Sparked by Spartan fears about the growth of Athenian power, the war involved most of the Greek world from Asia Minor to Sicily and from Byzantium (present-day Istanbul, Turkey) to Crete. It was a classic example of a war between a sea-power and a land-power. Athens controlled most of the Aegean and its coasts, and Sparta most of the Peloponnese and central Greece.

The end came with Lysander's destruction of the Athenian fleet at Aegospotami in 405 BC. After withstanding siege by both land and sea through the winter, Athens surrendered in 404 BC. As a result of this defeat, Athens' political power collapsed.

Peninsular War

A war of 1808–14 caused by the French emperor **Napoleon's** invasion of Portugal and Spain. British expeditionary forces under Sir Arthur Wellesley (Duke of **Wellington**), combined with Spanish and Portuguese resistance, succeeded in defeating the French at **Vimeiro** in 1808, **Talavera** in 1809, **Salamanca** in 1812, and **Vittoria** in 1813. The results were inconclusive, and the war was ended by Napoleon's forced abdication in 1814.

Pentagon

The headquarters of the US Department of Defence, in Arlington, Virginia, from 1947. Situated on the Potomac River opposite Washington, DC, the Pentagon is one of the world's largest office buildings (five storeys high and five-sided, with a pentagonal central court). It houses the administrative and command headquarters for the US armed forces and has become synonymous with the military establishment bureaucracy.

> The Pentagon building contains 27 km/ 17 mi of corridors, and possesses the largest private branch telephone exchange in the world.

Pershing, John Joseph (1860–1948)

US general after whom the Pershing surface-to-surface **missile** was named. Pershing served in the **Spanish–American War** in 1898, then in the

Philippines, and in Mexico in 1916–17. In **World War I**, he stuck to the principle of using US forces as a coherent formation, refusing to attach regiments or brigades to British or French divisions. He commanded the American Expeditionary Force sent to France in 1917–18.

In 1919 Pershing was made a full general of the US army, a rank that had been previously held by only four people: George Washington, Ulysses S Grant, William Tecumseh Sherman, and Philip Henry Sheridan.

Persian Wars

A series of conflicts between Greece and Persia in 499–479 BC. Persian involvement with Greece began when Cyrus (II) the Great (reigned 559–530 BC) conquered the Greek cities of western Asia Minor and ended with **Alexander** (III) **the Great's** conquest of Persia. The term 'Persian Wars' often refers just to the two Persian invasions of mainland Greece in 490 and 480/79 BC. The victory by Alexander marked the end of Persian domination of the ancient world and the beginning of Greek supremacy.

See also: *Salamis, Battle of.*

- The first Persian invasion was a seaborne expedition led by Darius (I) the Great. It was halted at the battle of Marathon in 490 BC.

- In 480 BC Darius' son, Xerxes I, invaded Greece by both land and sea. Eastern Greece as far south as Athens was overrun, but the Greek fleet defeated the Persians at Salamis, and a campaign in the vicinity of Plataea ended in a decisive Greek victory in 479 BC.

phalanx

A battle formation used in ancient Greece and Macedonia. It consisted of up to 16 lines of infantry with **spears** about 4 m/13 ft long, protected to the sides and rear by cavalry. The phalanx was deployed by Philip II and **Alexander the Great** of Macedon, and though more successful than the conventional hoplite (heavily armed infantry) formation it proved inferior to the Roman legion.

Philippine Sea, Battle of

A decisive **World War II** naval victory by the USA over Japan in June 1944. The last of the great carrier battles, it took place to the east of the Philippine islands and broke the back of the Japanese navy.

Plassey, Battle of

British victory under Robert **Clive** over the Nawab of Bengal, Suraj Dowla, on 23 June 1757. This brought Bengal under the control of the East India Company and hence under British rule. The battle took place at the former village of Plassey, about 150 km/95 mi north of Calcutta. Although outnumbered, Clive won the battle with minimal losses through Suraj's impetuous squandering of his advantage in an all-out bombardment that exhausted his ammunition. Clive used the support of his Indian banker allies to buy the defection of Suraj's general Mir Jafar, who he then installed as nawab.

The nawab's army included 35,000 foot soldiers, 18,000 cavalry, and 50 guns. Clive's forces numbered just 3,000, of which he lost 72.

platoon

The smallest infantry subunit of an army. It contains 30–40 soldiers and is commanded by a lieutenant or second lieutenant. There are three or four platoons in a **company**.

Poitiers, Battle of

A battle between the English and the French that took place outside Poitiers on 13 September 1356 during the **Hundred Years' War**. It ended in victory for Edward the Black Prince over King John II of France. King John, his son Philip, and 2,000 knights were taken prisoner, and about 3,000 French were killed. English losses were small, and Edward was able to make a gradual retreat to Bordeaux.

Poltava, Battle of

A battle that took place in 1709 outside the Ukrainian city of Poltava, 300 km/186 mi southeast of Kiev; in which Peter the Great routed a Swedish and Ukrainian Cossack force led by Charles XII of Sweden. This victory ended the **Great Northern War**.

Port Arthur, Battle of

A victory by the Japanese over the Russians during the Russo–Japanese War, after besieging the city of Port Arthur

Gen Stössel, who surrendered Port Arthur to the Japanese, was an incompetent who had been relieved of his post by the tsar, but concealed the order from his staff and continued to exercise command.

in Manchuria (now Lüshun, China) from May 1904 to January 1905. The Russian occupation of Port Arthur in 1897, formalized as a lease in 1898, had been one of the main points of contention, and its loss was a significant blow to Russian morale.

prisoner of war (POW)

Person captured in war that has fallen into the hands of, or surrendered to, an opponent. Such captives may be held in prisoner-of-war camps. The treatment of POWs is governed by the **Geneva Convention**.

Punic Wars

Three wars fought between Rome and Carthage in the 3rd and 2nd centuries BC.

- First Punic War (264–241 BC) resulted in the defeat of the Carthaginians under Hamilcar Barca and the cession of Sicily to Rome.
- Second Punic War (218–201 BC). Hannibal invaded Italy, defeated the Romans at Trebia, Trasimene, and Cannae, but was finally defeated by Scipio Africanus Major at Zama (now in Algeria).
- Third Punic War (149–146 BC) ended in the destruction of Carthage, and its possessions becoming the Roman province of Africa.

Q-Boats also called mystery ships
Small freighters with guns concealed in a collapsible deck structure used by the British in **World War I** to trap German submarines.
- On being hailed by a U-boat, a 'panic party' would hastily abandon the ship by lifeboat, leaving a fighting party concealed on board.
- The U-boat would be lured into sailing closer to the seemingly abandoned ship, when suddenly the ship's guns would be revealed and open fire.
- Several U-boats were sunk by this ploy, but it soon became well known and was abandoned.

Québec, Battle of
A battle fought in Canada on the Plains of Abraham, a plateau near Québec, between the British and French in 1759 during the **Seven Years' War**. The British, under Gen **Wolfe**, defeated French troops led by Montcalm, but both commanders died in the fighting. The capture of Québec established British supremacy in Canada.

> The Plains of Abraham lie southwest of the city in the National Battlefields Park, which contains a monument to Gen Wolfe. Also to the southwest is Wolfe's Cove, traditionally the landing point of Wolfe's troops in 1759.

Quiberon Bay, Battle of
A naval battle off the northwest coast of France in 1759 during the **Seven Years' War**. In Quiberon Bay, the British admiral Hawke defeated a French fleet under the Marquis de Conflans, wrecking French plans for an invasion of England and reasserting British command of the sea.

Quisling, Vidkun Abraham Lauritz Jonsson (1887–1945)
Norwegian politician whose name became synonymous with 'traitor'. Leader of the Norwegian Fascist Party from 1933, Quisling aided the Nazi

invasion of Norway in 1940, during **World War II**, by delaying mobilization and urging non-resistance. He was made premier by Hitler in 1942, and was arrested and shot as a traitor by the Norwegians in 1945.

radar
Acronym for a device for locating objects in space, direction finding, and navigation by means of transmitted and reflected high-frequency radio waves. It is widely used in warfare to detect enemy aircraft and missiles.

- Radar proved invaluable in the **Battle of Britain** 1940, when the ability to spot incoming German aircraft did away with the need to fly standing patrols.
- Chains of ground radar stations are used to warn of enemy attack in the North Warning System, consisting of 52 stations across the Canadian Arctic and northern Alaska.
- To avoid detection, various devices, such as modified shapes (to reduce their radar cross-section), radar-absorbent paints, and electronic jamming are used.
- To pinpoint small targets laser 'radar', instead of microwaves, has been developed.

RAF
Abbreviation for **Royal Air Force**

Ramillies, Battle of
A victory for English and Dutch troops, led by the Duke of **Marlborough**, over the French near Ramillies, 19 km/12 mi north of Namur, in Belgium. The battle took place on 23 May 1706 during the **War of the Spanish Succession**. The French lost all their artillery and suffered some 15,000 casualties; English and Dutch losses were fewer than 4,000.

Rangers
Specially trained, specially equipped, highly mobile troops deployed by the US Army in **World War II** and modelled on the British **commandos**.

recoilless gun
A rifle that directs some of the explosion of the propellant cartridge backwards, balancing the recoil caused by the ejection of the projectile.

A recoilless gun saves weight in the construction of the mounting. Its principal defects are its short range and a prominent backblast, both due to the propellant gases being ejected to the rear.

reconnaissance
The gathering of information by the military about an objective. It can be carried out by a reconnaissance (recce) patrol or from a small, fast-moving vehicle, or an aircraft configured for reconnaissance.

Red Army
The army of the former USSR until 1946, when it became known as the Soviet Army. Founded by the Russian revolutionary Leon Trotsky, the Red Army developed from the Red Guards, volunteers who were in the vanguard of the Bolshevik revolution of 1917. It took its name from its rallying banner, the red flag. At its peak, during **World War II**, the Red Army included about 12 million men and women.

The revolutionary army that helped the communists under **Mao Zedong** win power in China in 1949 was also popularly known as the Red Army.

Red Cross
International relief agency founded by the **Geneva Convention** of 1863 as the International Federation of the Red Cross. It was instigated by the Swiss doctor Henri Dunant to assist the wounded and **prisoners of war**. Its symbol is a symmetrical red cross on a white ground. (Muslim countries use the red crescent symbol instead).

In December 1996 six foreign-aid workers, including five nurses, were shot dead by masked guerrillas at a Red Cross hospital in Chechnya in the former Soviet Union. The incident has spurred efforts to draw up new rules to protect vulnerable people working for humanitarian organizations, such as the Red Cross, in conflict zones.

The Red Cross also deals with other problems of war, such as refugees and the care of the disabled, as well as care of victims of natural disasters such as floods and earthquakes. It was awarded the Nobel Peace Prize 1917 and 1944.

redoubt
In trench warfare, a small, enclosed, defensive trench work employed in conjunction with a system of infantry trenches. It can form a strong point of resistance even after the rest of the trenches have been destroyed or captured.

regiment
Military formation equivalent to a **battalion** in parts of the British army, and equivalent to a **brigade** in the armies of many other countries.

> **VARIATIONS**
> - Tank, engineer, artillery, and some logistic units in the British army use the term 'regiment' synonymously with 'battalion'.
> - In the British infantry, a regiment may include more than one battalion, and soldiers belong to the same regiment throughout their career.

Remagen
German town on the left bank of the River Rhine, 25 km/15 mi southeast of Bonn. It was a crucial point in the Allied crossing of the river for the advance into Germany at the end of **World War II**.

On 7 March 1945 a patrol of the US 9th Armored Division discovered that the Ludendorff Bridge crossing the Rhine at Remagen had escaped destruction. Calling up troops of 87th Division, they seized the bridge, giving them a valuable crossing point.

The bridge had been bombed earlier in the war by the Allies, and was now bombed by the Germans. This, coupled with the heavy traffic of tanks across it, caused the bridge to collapse some days later, but by that time other crossings had been secured.

remotely piloted vehicle (RPV)
A crewless mini-aircraft used for military surveillance and to select targets in battle. RPVs barely show up on radar, so they can fly over a battlefield without being shot

> RPVs were used by Israeli forces in 1982 in Lebanon and by the Allies in the 1991 Gulf War.

down, and are equipped to transmit TV images to an operator on the ground. The US system is called Aquila and the British system Phoenix.

revolver
Pistol with a revolving magazine that carries the bullets and their charges, enabling five or six shots to be taken in succession before reloading.

Richthofen, Manfred, Freiherr von (1892–1918)
German aviator in **World War I** known as the 'Red Baron'. He commanded the 11th Chasing Squadron, known as Richthofen's Flying Circus, and shot down 80 aircraft before being killed in action.

THE RED BARON

- Originally a cavalryman (Lancer), von Richthofen transferred to the air corps and eventually became the most famous 'ace' of the German service.
- He scored his 80th victory in April 1918, was shot down behind British lines on the Somme in the same month, and was buried with full military honours.

Rickover, Hyman George (1900–1986)
Russian-born US naval officer who was responsible for the development of the first nuclear submarine, the *Nautilus*, in 1954. During **World War II**, he worked on the atomic-bomb project, headed the navy's nuclear reactor division, and served on the Atomic Energy Commission. He was promoted to the rank of admiral in 1973.

After retiring in 1982, Rickover became an outspoken critic of the dangers of nuclear research and development.

rifle
Firearm that has spiral grooves (rifling) in its barrel. When a bullet is fired, the rifling makes it spin, thereby improving accuracy. Rifles were first introduced in the late 18th century.
See also: *Lee-Enfield and Mauser rifles.*

River Plate, Battle of the

A **World War II** naval battle fought in December 1939 in the South Atlantic, between a British cruiser squadron of three ships and the German 'pocket battleship' *Graf Spee*.

Although the British cruisers were no match for the battleship, they did sufficient damage to make the German ship's captain break off and run for shelter in Montevideo, Uruguay. The British followed, and waited in international waters outside the neutral port. The Uruguay government ordered the Germans to leave after 72 hours. Hitler, reluctant to risk the *Graf Spee* being sunk by the heavier British warships that were sailing for the River Plate, ordered the captain to scuttle the vessel.

RN

Abbreviation for **Royal Navy**.

Roberts, Frederick Sleigh, 1st Earl Roberts (1832–1914)

British field marshal. During the Afghan War of 1878–80 he occupied Kabul, and during the Second South African War from 1899 to 1902 he made possible the annexation of the Transvaal and Orange Free State.

Born in India, Roberts joined the Bengal Artillery 1851, and served through the Indian Mutiny of 1857–58 (receiving the VC), and the Abyssinian campaign of 1867–68. After serving in Afghanistan and making a victorious march to Kandahar, he became commander-in-chief in India in 1885–93 and later in Ireland. In 1899 he received the command in South Africa, where he occupied Bloemfontein and Pretoria. He was made Earl in 1900.

rocket

Projectile driven by the reaction of gases produced by a fast-burning fuel. Unlike jet engines, which are also reaction engines, modern rockets carry their own oxygen supply to burn their fuel and do not require any surrounding atmosphere. For warfare, rocket heads carry an explosive device.

The development of rockets as a means of propulsion to high altitudes, carrying payloads, started in the interwar years with the state-supported work in Germany (primarily by German-born US rocket engineer Wernher von Braun) and the work of Robert Hutchings Goddard (1882–1945) in the USA. Being the only form of propulsion available that can function in a vacuum, rockets are essential to exploration in outer space.

See also: *nuclear warfare; missile.*

Rommel, Erwin Johannes Eugen (1891–1944)

World War II field marshal nicknamed 'Desert Fox'. He served in **World War I**, and in World War II played an important part in the invasions of central Europe and France. He was commander of the North African offensive from 1941, when he was defeated in the Battles of El **Alamein** and, in March 1943, expelled from Africa.

Rommel was commander-in-chief for a short time against the Allies in Europe in 1944 but, as a sympathizer with the Stauffenberg plot to assassinate Hitler, he was forced to commit suicide.

Rorke's Drift, Battle of

A British victory over a Zulu army on 22 January 1879 at Rorke's Drift, a farm about 170 km/105 mi north of Durban, during the Anglo-Zulu War in South Africa. A small British force on the farm, which was little more than a field hospital, held off 4,000 Zulus who had just defeated a much larger British force at Isandhlwana. Casualties amounted to 17 British killed, in comparison to 400 Zulus.

At 7.30 a.m. on 23 January the Zulus reappeared, but they simply sat down on a nearby hill and watched for about an hour before they stood, turned about, and went. Both sides had simply had enough and the battle died out.

Roses, Wars of the

Intermittent civil strife in England from the 1450s to 1487. Failure in the **Hundred Years' War** and incompetent government at home created opposition to the court circle around Henry VI. The opposition was led by Richard, Duke of York, who eventually claimed the throne for himself. Richard was killed in battle at Wakefield in 1460, but his son won the crown as Edward IV. Fighting recurred in 1470–71, when the Lancastrians invaded and were beaten at Barnet and Tewkesbury, and in 1483–87, when first Richard III and then Henry VII usurped the throne.(See map on p.154)

- The name 'Wars of the Roses' is a misleading 19th-century invention.
- Using a red (Lancaster) and a white (York) rose as symbols of the conflict began early, but the idea that the late 15th century was a period of unremitting conflict only brought to an end by Henry VII was a myth.

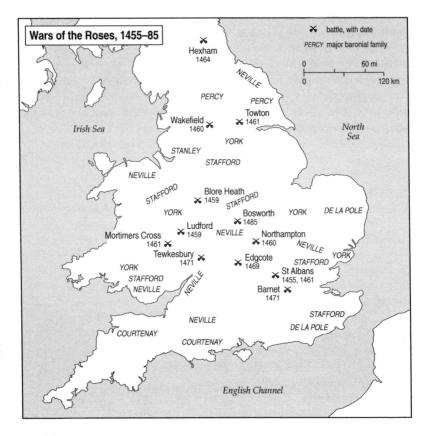

Royal Air Force (RAF)

The **air force** of Britain. The RAF was formed in 1918 by the merger of the Royal Naval Air Service and the Royal Flying Corps.

Royal Marines

British military force trained for amphibious warfare. See **marines**.

Royal Navy (RN)

Britain's **navy**. established by Alfred the Great in the 9th century. It gained the title Royal Navy in the reign of Charles II (1660–85), and was the means by which the British Empire extended itself around the world from the 17th century. It played a key role in Britain's stand against **Napoleon** and was never again challenged by the French after **Trafalgar** in 1805.

The Royal Navy was the world's most powerful navy until well into the 20th century. In **World War I** its main task was to protect shipping from submarine attack. After **World War II** the Royal Navy was second in size only to the US Navy but continued to be a world leader, especially in submarine warfare. By the 1980s, as a result of defence cuts, the Royal Navy had declined to third in world size, after the US and Soviet navies.

NUCLEAR NAVY

- The Royal Navy has been responsible for Britain's nuclear deterrence from 1969 and in 1995 had a fleet of four nuclear submarines.
- As a fighting force in recent times, the Royal Navy played a vital national role in the Falklands War in 1982, and also formed part of an international force in the Korean War (1950–53), Gulf War (1990–91), and Balkans War (1992–95).

Russo–Japanese War

War between Russia and Japan in 1904–05, which arose from conflicting ambitions in Korea and Manchuria, specifically, the Russian occupation of **Port Arthur** (modern Lüshun) in 1897 and the Amur province in 1900. Japan captured Port Arthur in January 1905, took **Mukden** (modern Shenyang) on 10 March, and on 27 May defeated the Russian Baltic fleet, which had sailed halfway around the world to the Tsushima Strait. A peace was signed on 23 August 1905. Russia surrendered its lease on Port Arthur, ceded southern Sakhalin to Japan, evacuated Manchuria, and recognized Japan's interests in Korea.

Saipan
An island of the Marianas group, about 1,900 km/1,200 mi north of New Guinea, that was occupied by the Japanese in **World War II**. When US troops recaptured it in June 1944, several hundred Japanese civilians committed mass suicide rather than be taken prisoner.

Saladin or Salah al-Din Yusuf ibn Ayyub (c. 1138–1193)
Kurdish conqueror of the Kingdom of Jerusalem, who was renowned for knightly courtesy. Saladin was tutored in the military arts by his uncle, a general in Aleppo, before becoming the ruler of Egypt in 1169 and Aleppo in 1183. He recovered Jerusalem from the Christians in 1187, and captured nearly all the castles of the Christian Kingdom by 1189, precipitating the Third **Crusade**. He was besieged at Acre, defeated by Richard (I) the Lionheart at Arsuf in 1191, and made peace with Richard in 1192.

Saladin lived to see his greatest opponent, Richard the Lionheart, leave for Europe without having recaptured Jerusalem, so proving himself to be the superior strategist of the two.

Salamanca, Battle of
A battle of the **Peninsular War** in which the British led by the Duke of **Wellington** scored a victory over the French under Marshal Auguste Marmont. It took place on 22 July 1812 near Salamanca, 170 km/105 mi northwest of Madrid.

Wellington entered Spain with 42,000 troops and took up a position south of Salamanca, which protected his supply route back to Portugal. The 46,000 French troops occupying Salamanca moved out to oust the British and capture the road. Wellington forestalled their plans by sending a smaller force to make a head-on attack on the advancing French army.

Once this battle was under way, he sent the rest of his infantry to attack the French flank. Heavy fighting ensued, and Wellington then launched his cavalry against the other flank of the French column. This attack scattered the French and totally ended their resistance. French casualties came to 15,000, while British losses were 6,000.

Salamis, Battle of

A sea battle fought in the Strait of Salamis, west of Athens, Greece, in 480 BC between the Greeks and the invading Persians (see **Persian Wars**). Despite being heavily outnumbered, the Greeks inflicted a crushing defeat on the enemy.

Tradition says that Themistocles, the Athenian commander of the Greek fleet of 370 galleys, sent the Persians a fake message, ostensibly from a spy, advising that the Greek fleet was about to withdraw and the Persians should blockade the entrance to the Bay of Eleusis. The Persians fell for the ruse and spread their 1,000 ships thinly across the bay. The Greeks came out into the bay at full speed, broke the Persian line, and created mayhem in all directions. Some 500 Persian ships were lost, but only 40 Greek galleys.

PRIDE BEFORE A FALL

- The Persians were so confident they could deal with a mere 370 vessels that they had a throne prepared for their king, Xerxes, on nearby Mount Aegaleus to give him a grandstand view.
- Disgusted at the humiliating defeat, Xerxes returned to Asia leaving a subordinate, Mardonius, to continue the land campaign.

SALT

Abbreviation for **Strategic Arms Limitation Talks**, a series of US–Soviet negotiations between 1969 and 1979.

samurai, or bushi

Japanese term for the warrior class that became the ruling military elite for almost 700 years. A samurai (Japanese 'one who serves') was an armed retainer of a *daimyo* (large landowner), with specific duties and privileges and a strict code of honour. The system was abolished in 1869 and the government pensioned off all samurai.

- From the 16th century, commoners were not allowed to carry swords, whereas samurai had two swords, and the higher class of samurai were permitted to fight on horseback.

- After the Meiji restoration, the introduction of universal conscription in 1872 ended the samurai's military role, and many of them rebelled.
- Their last uprising was the Satsuma Rebellion of 1877–78, in which 40,000 samurai took part.

Saratoga, Battle of

A battle of the **American Revolution** that took place in September– October 1777 near Saratoga Springs, about 240 km/150 mi north of New York, and ended in a humiliating defeat for the British.

On 17 October Gen John Burgoyne surrendered to the American commander Gen Horatio Gates. Under the terms of the surrender, known as the Convention of Saratoga, the British were to be allowed to march to Boston and there embark for England, but Congress refused to ratify it and Burgoyne and his force became prisoners of war until peace was signed.

SAS, Abbreviation of Special Air Service

A division of the British Army specially trained in carrying out operations behind enemy lines. It was started in North Africa in 1942, during **World War II**, by Sir David Stirling (1916–90) and is the most secretive of all the British forces. After the war, the SAS saw action in trouble spots where Britain was involved, such as Malaya (now Malaysia) and Aden (now in the Republic of Yemen), and more recently in Northern Ireland.

Who dares wins

The **SAS** motto

SBS

Abbreviation for Special Boat Service, the British **Royal Navy's** equivalent of the **SAS** (Special Air Service).

Scapa Flow

A large, protected sea area in the Orkney Islands, Scotland, between Mainland, Flotta, South Ronaldsay, and Hoy, and until 1957 a base of the Royal Navy. It was the main base of the Grand Fleet during **World War I** and in

A German U-boat penetrated the anchorage on 14 October 1939, and sank the battleship Royal Oak with the loss of 810 men.

1919 was the scene of the scuttling of 74 surrendered German warships (62 of which have since been salvaged). Reactivated as a base in **World War II**, today it is an attraction for scuba divers.

schnorkel

A device that permits submarines to draw air from, and emit fumes to, the surface while running their engines and remaining underwater. Invented in 1938 by a Dutch naval engineer and fitted to some Dutch submarines, it was adopted by the Germans in 1942–43, during **World War II**, to allow their submarines to run on diesels and charge batteries whilst remaining submerged, so as to avoid detection by **radar**.

Schwarzkopf, Norman (1934–)

US general, supreme commander of the Allied forces in the **Gulf War** of 1991. Nicknamed 'Stormin' Norman', he planned and executed a blitzkrieg campaign, 'Desert Storm', sustaining remarkably few Allied casualties in driving the Iraqi army out of Kuwait.

A graduate of the military academy at West Point, he became an infantry soldier and later a paratrooper. He was a battalion commander in the Vietnam War in 1969–70 and deputy commander of the US invasion of Grenada in 1983. His success in the Gulf War made him a popular hero in the USA. He retired from the army in August 1991.

> He is neither a strategist, nor is he schooled in the operational art, nor is he a tactician, nor is he a general, nor is he a soldier. Other than that, he is a great military man.
>
> **Norman Schwarzkopf**, US general, on the Iraqi president Saddam Hussein, March 1991.

SDI

Abbreviation for **Strategic Defence Initiative**.

Sedan, Battle of

A disastrous French defeat by the Prussians during the **Franco-Prussian War** on 2 September 1870 at Sedan, a fortified town in northern France, close to the Belgian border, about 195 km/120 mi northeast of Paris. The victory cost the Prussians some 9,000 casualties, but the French lost 17,000 killed and wounded and 104,000 prisoners of war, including Emperor Napoleon III. The French were left with no effective regular army – thereafter war would have to be carried on by mostly citizen armies – and only Paris still held out. In the aftermath of the defeat, the French decided to return to republican government.

self-propelled artillery

A class of **armoured fighting vehicle** that mounts a heavy gun or missiles, moves under its own power, usually on tracks, and provides fire support in forward areas of the battlefield.

- The first British self-propelled guns were produced in the second half of 1916, and were known as Gun Carriers Mark I.
- During **World War II** self-propelled artillery was used extensively by the Germans, who mounted large-calibre guns on tank chassis as field artillery, and high velocity guns on self-propelled mountings as tank destroyers.

semaphore

A visual signalling code in which the relative positions of two moveable pointers or hand-held flags stand for different letters or numbers. The system is used by ships at sea and for railway signals.

Semtex

Plastic explosive manufactured in the Czech Republic. It is safe to handle (it can only be ignited by a detonator) and difficult to trace, since it has no smell. It has been used by guerrillas in the Middle East and by the **IRA** in Northern Ireland.

A quantity of 0.5 kg/1.1 lb of Semtex is thought to have been the cause of an explosion that destroyed a Pan-American Boeing 747 airliner over Lockerbie, Scotland, in December 1988, killing 270 people.

services, armed

See **armed forces**.

Sevastopol, Siege of

A successful British and French siege of Sevastopol, a fortified Russian town on the Black Sea, from October 1854 to September 1855 during the

Crimean War. The Russian fleet was based in Sevastopol harbour, so the town was the prime objective of the main Allied attack in the Crimea.

On 7 June 1855 the French took Mamelon fortress, which protected the Malakoff line, the principal defensive line for Sevastopol. A regular bombardment of the town began early in August and the French victory at Chernaya signalled the end of a lengthy Russian resistance. After a final assault by the Allies on 5 September, Sevastopol was evacuated and the Russians retreated inland, leaving their wounded behind.

Seven Years' War

A war (Known as the French and Indian War in North America) in 1756–63 that arose from the conflict between Austria and Prussia, and between France and Britain, over colonial supremacy. Britain and Prussia defeated France, Austria, Spain, and Russia. Britain gained control of India and many of France's colonies, including Canada. Spain ceded Florida to Britain in exchange for Cuba. Fighting against great odds, Prussia eventually established itself as one of the great European powers. The war ended with the Treaty of Paris 1763, signed by Britain, France, and Spain.

Sherman, Thomas West (1813–1879)

A US soldier who fought in Florida, the Mexican War, and on the frontier. A stern disciplinarian, he commanded the Port Royal, South Carolina, expedition in 1861 and afterward fought in Mississippi and Louisiana. He lost a leg in action at Port Hudson in 1863.

Sherman, William Tecumseh (1820–1891)

Union General in the **American Civil War**. Born in Ohio, the son of a judge, Sherman graduated from West Point military academy in 1840. He saw action at **Bull Run** in 1861, served under Gen **Grant** at the Battle of **Shiloh** in 1862 and at **Vicksburg** in 1863. He captured and burned Atlanta, Georgia, in 1864, and then began his 'march to the sea' across Georgia, stripping houses and burning crops to demoralize the enemy. He marched through South Carolina, burned the state capital Columbia, and in 1865 forced the Confederate general Johnson in North Carolina to surrender.

shield

A means of protection in combat used from earliest times till the 14th century, when **firearms** were invented, rendering the shield useless. Shields down the ages varied in shape and size and were made of many different materials.

shield *Assault on a medieval castle using a protective formation of shields.*

SHIELD SUCCESSION

- The Romans used a large, rectangular wooden shield covered in leather.
- In the Middle Ages, knights wearing helmets that covered the face were identified by the coats of arms on their shields, which were usually small and triangular.
- Shields made of synthetic materials are still used by police across the world in riot-control situations.

Shiloh, Battle of

In the **American Civil War**, a Confederate defeat by Union forces under Gen Ulysses S **Grant** on 6–7 April 1862 near Shiloh Church, about 150 km/95 mi east of Memphis, Tennessee. The Confederate failure to stop Grant's army gave the Union control of the Mississippi valley. Confederate losses at Shiloh were 10,690 killed or wounded and 960 taken prisoner. The Union side lost 10,150 killed or wounded and 4,045 prisoners.

shrapnel

Artillery projectile consisting of a hollow shell loaded with lead or steel balls and a small charge of gunpowder. When the shell is fired over enemy lines, the charge is ignited by a time fuse, ejecting the balls forward like a shotgun blast and spraying bullets down onto the troops below.

An efficient killer when used against troops in the open, shrapnel lost its efficacy when troops began to entrench and take cover, and was gradually superseded by high-explosive shells.

- Shrapnel is named after its inventor, the English artillery officer Henry Shrapnel (1761–1842).
- The term is now also used to refer to fragments of metal produced from an exploding device.

siege

Prolonged assault on a fortified position, often involving a blockade, aimed at forcing the enemy to surrender.

Down the ages, siege techniques have evolved along with fortifications and warfare technology. For example, stone walls, which could withstand attack by giant catapults, battering rams, and siege engines (movable wooden towers filled with armed men), were later breached by **cannon**; the first use of the tank, in **World War I**, exposed the vulnerability of earth trench defences.

siege *Fortified town under attack in the 14th century.*

See also: *the Sieges of Acre; Badajoz; Gibraltar; Ladysmith; Leningrad; Leyden; Lucknow; Madrid; Mafeking; Paris; Sevastopol; Stalingrad; Vienna.*

Siegfried Line

In **World War I**, a defensive line established in 1917 by the Germans in France; in **World War II** a subdivision of the main **Hindenburg Line**; the Allies' name for the West Wall, a German defensive line established along its western frontier, from the Netherlands to Switzerland.

Sinai, Battle of

A tank battle from 6 to 24 October 1973 during the **Yom Kippur War** between Israel and

The Battle of Sinai was one of the longest tank battles in history.

Egypt. On 16 October Israeli troops crossed the Suez Canal, cutting off the Egyptian 3rd Army.

Singapore, fall of

Capture, in 1942, of the island of Singapore, off the tip of the Malay Peninsula, by the Japanese in **World War II**. Before the war, Singapore, then a British colony (today an independent country) had been developed as a naval base from which a powerful fleet could operate, and was protected against sea attack by fixed coastal defences. However, defence along the Malay Peninsula proved useless when the Japanese pushed the British forces down the Peninsula into Singapore in 1942, and on 15 February the British commander surrendered. The British retook Singapore on 5 September 1945.

Sino–Japanese Wars

Two wars waged by Japan against China to expand to the Asian mainland.

First Sino–Japanese War 1894–95

Under the treaty of Shimonoseki, Japan secured the 'independence' of Korea, cession of Taiwan and the nearby Pescadores Islands, and the Liaodong peninsula (for a naval base). France, Germany, and Russia pressured Japan into returning the last-named, which Russia occupied 1896 to establish Port Arthur (now Lüshun); this led to the **Russo–Japanese War** of 1904–05.

Second Sino–Japanese War 1931–45

1931–32 The Japanese occupied Manchuria, which they formed into the puppet state of Manchukuo.

1937–38 Chinese leaders Jiang Jie Shi (Chiang Kai-shek) and **Mao Zedong** allied to fight the Japanese; war was renewed as the Japanese overran northeastern China and seized Shanghai and Nanjing, and in 1938 Wuhan and Guangzhou.

1941 Japanese attack on the USA (*see* **Pearl Harbor**) led to the extension of lend-lease aid to China and US entry into war against Japan and its allies.

1945 The Chinese received the Japanese surrender at Nanjing in September, after the end of **World War II**.

Six-Day War

Another name for the third **Arab–Israeli War**.

SLBM
Abbreviation for **submarine-launched ballistic missile**; *see* **nuclear warfare**.

Slim, William Joseph, 1st Viscount Slim (1891–1970)
British field marshal who was knighted for his service in **World War II**. Slim served in the North Africa campaign in 1941 and then commanded the 1st Burma Corps in 1942–45, stemming the Japanese invasion of India and forcing the Japanese out of Burma (now Myanmar). He was governor general of Australia from 1953 to 1960. A KCB since 1944, he was made Viscount in 1960.

> In a battle nothing is ever as good or as bad as the first reports of excited men would have it.
>
> **William Slim**, *Unofficial History*.

smart weapon
Programmable bomb or missile that can be guided to its target by laser technology, TV homing technology, or terrain-contour matching (TERCOM). A smart weapon relies on its pinpoint accuracy to destroy a target rather than on the size of its warhead. Examples are the **cruise missile** (Tomahawk), laser-guided artillery shells (Copperhead), laser-guided bombs, and short-range TV-guided missiles (SLAM).

Smart weapons were first used on the battlefield in the Gulf War 1991, but only 3% of all the bombs dropped or missiles fired were smart. Of that 3%, it was estimated that 50–70% hit their targets, which is a high accuracy rate.

smoke screen
In warfare, cloud of smoke released by means of smoke shells or chemical generators. It drifts in the wind and obscures the view of the enemy so as to conceal movement of troops or ships.

sniper
Soldier trained in accurate long-range shooting, scouting, and reconnaissance.

The sniper's function is to kill individual 'high-value' targets such as officers and specialists, to observe the enemy's activities and report upon them, and to seek out and kill enemy snipers.

SOE
Abbreviation for **Special Operations Executive**.

Somme, Battle of the
Allied offensive in **World War I** during July–November 1916 on the River Somme in northern France. It was planned by the Marshal of France, Joseph Joffre, and British commander-in-chief Douglas **Haig**.

The German offensive around St Quentin of March–April 1918 is sometimes called the Second Battle of the Somme.

The first Battle of the Somme was launched on 1 July by British and French troops against well-entrenched German dugout positions.

- The British suffered the heaviest casualties in their history; 19,000 men were killed on the first day.
- On 15 September tanks were used for the first time; some 47 tanks were available to the Allies of which most broke down.
- When the battle finally died away in mid-November the total Allied gain was about 13 km/8 mi at a cost of 615,000 Allied and about 500,000 German casualties.

South African Wars
Two wars between the British and the Dutch settlers (Boers) in South Africa, essentially over gold and diamond deposits in the Transvaal.

- The War of 1881 started after the Boers in the Transvaal reasserted the independence they had surrendered in 1877 in return for British support against the Africans. The British were defeated at Manjuba, and the Transvaal became independent.
- The War of 1899–1902, also known as the Boer War, began after the Cape Colony prime minister, Cecil Rhodes, attempted to instigate a revolt among the uitlanders (non-Boer immigrants) against Kruger, the Transvaal president. The Boers invaded British territory, besieging Ladysmith, Mafeking, and Kimberley, but eventually conceded defeat by the Peace of Vereeniging.
- British commander **Kitchener** countered Boer guerrilla warfare by putting the noncombatants who supported them into concentration camps, where about 26,000 women and children died of sickness.

> When is a war not a war? When it is carried on by methods of barbarism.

Henry Campbell-Bannerman, British Liberal politician, speech condemning the Boer War (one of the South Africa Wars) June 1901.

Spanish–American War
A brief war in 1898 between Spain and the USA over Spanish rule in Cuba and the Philippines. The complete defeat of Spain made the USA a colonial power. The Treaty of Paris ceded the Philippines, Guam, and Puerto Rico to the USA. Cuba became independent. The USA paid $20 million to Spain to end Spain's colonial presence in the Americas.

Spanish Armada
Fleet sent by Philip II of Spain against England in 1588. Consisting of 130 ships, it sailed from Lisbon and carried on a running fight up the Channel with the English fleet of 197 small ships under Howard of Effingham and Francis **Drake**. The Armada anchored off Calais but fire ships forced it to put to sea, and a general action followed off Gravelines. What remained of the Armada after the skirmish, escaped around the north of Scotland and west of Ireland, suffering many losses by storm and shipwreck on the way, only about half the original fleet returned to Spain.

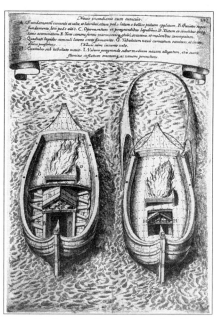

Spanish Armada *Contemporary drawings of fire ships launched against the Armada.*

Spanish Civil War
See **Civil War, Spanish.**

spear
A long pole or shaft ending in a sharp point, or with a pointed head, one of the oldest weapons. The heads were made of bone, stone, bronze, or iron, depending on the period. Spears were made for throwing at the enemy from a distance or for close-quarter stabbing. Their use in close combat declined in the 18th century after the invention of the **bayonet**.

Lances, carried by medieval knights, and pikes, used by infantry during the Renaissance, were variations on the spear.

See also: *phalanx*.

Special Operations Executive (SOE)
British intelligence organization established in June 1940 to gather intelligence and carry out sabotage missions inside German-occupied Europe during **World War II**.

Some 11,000 SOE agents were eventually employed, but screening was careless and a number of German agents infiltrated the organization, fatally damaging many operations before they were detected and removed.

SS
Abbreviation for the German Schutz-Staffel ('protective squadron'), a **Nazi** elite corps established in 1925. Under Himmler its 500,000 membership included the full-time Waffen-SS (armed SS), who fought in **World War II**, and part-time members. The SS performed state police duties and was brutal in its treatment of the Jews and others in the concentration camps and occupied territories. It was condemned as an illegal organization at the Nuremberg Trials of war criminals (1945–46).

Stalingrad, Siege of
German siege of former Soviet city of Stalingrad (now Volgograd, in Russia) between August 1942 and January 1943 in **World War II**. The Siege of Stalingrad was a horrific campaign, with both sides sustaining heavy casualties and the Germans finally being driven out.

- The German 6th Army, reinforced by the 4th Panzer Army, launched the first major assault on 19 August 1942.
- The initial advance through the suburbs was relatively smooth, but once into the built-up areas it became a house-to-house battle, which went on for two months.

- Meanwhile the Soviets were preparing a massive counterattack with 1 million troops, 13,500 guns, and 894 tanks commanded by Marshal Georgi Zhukov. This was launched on 19 November, sweeping around the flanks of the 6th Army, encircling, and destroying it.
- The 6th Army surrendered on 31 January 1943 with the loss of 1.5 million troops, 3,500 tanks, 12,000 guns and mortars, 75,000 vehicles, and 3,000 aircraft.
- There were 750,000 Soviet military casualties and an unknown number of civilian deaths.

Stamford Bridge, Battle of

Battle on 25 September 1066 at Stamford Bridge, northeast of York, England, in which King Harold II defeated and killed Harold Hardraada, King of Norway.

Harold was in the south with an army he had collected to meet the anticipated invasion by the Normans, and upon news of the Norse invasion immediately marched north. He confronted the Norse army at Stamford Bridge and a fierce battle ensued, in which both the Norwegian king and Tostig, the English king's exiled brother, were killed.

On meeting the Norse at Stamford Bridge, Harold offered the Norse king generous compensation if he retired or seven feet of earth for a grave if he stayed!

A few days later, news came that William the Conqueror had landed at Pevensey; Harold marched south and with a weary army fought the Battle of **Hastings**.

Star Wars

Popular term for the **Strategic Defense Initiative** announced by US president Ronald Reagan in 1983.

stealth technology

Methods used to make an aircraft invisible to radar and to detection by visual means and heat sensors. This is achieved by a combination of aircraft-design elements: smoothing off all radar-

The US F-117A stealth fighter-bomber was used successfully during the 1991 **Gulf War** to attack targets in Baghdad completely undetected.

reflecting sharp edges, covering the aircraft with radar-absorbent materials, fitting engine coverings that hide the exhaust and heat signatures of the aircraft, and other, secret technologies.

strategic bombing

Bombing of enemy territory with the aim of disrupting its economy and destroying morale. During **World War II** strategic bombing of cities was extensively used by both the **Axis** and the Allied powers. Rotterdam in the Netherlands, Warsaw in Poland, and Dresden in Germany all experienced heavy bombardment.

Strategic Defense Initiative (SDI)

Popularly known as Star Wars, an attempt by the USA to develop a defence system, based in part outside the Earth's atmosphere, against incoming nuclear missiles. It was announced by US president Ronald Reagan in March 1983, and the research had by 1990 cost over $16.5 billion.

SDI was scaled down in 1991 and an increased emphasis placed on limited defence, resulting in Global Protection Against Limited Strikes (GPALS), a programme that would be capable of destroying only a few missiles aimed at the USA and its allies.

The Clinton administration in 1993 renamed the SDI Organization the Ballistic Missile Defence Organization to reflect its focus on defence against short-range, rather than long-range strategic missiles.

- In 1988, the US Joint Chiefs of Staff announced that they expected to be able to intercept no more than 30% of incoming missiles.
- By 1996, plans to deploy an initial ground-based system of 100 interceptors capable of shooting down incoming missiles replaced the space-based system originally conceived by Reagan.

submarine

Underwater warship. The first underwater boat was constructed in 1620 for James I of England by the Dutch scientist Cornelius van Drebbel (1572–1633). A naval submarine, or submersible torpedo boat, the *Gymnote*, was launched by France in 1888. Submarine warfare was established as an effective form of naval tactics in **World Wars I** and **II**.

The conventional submarine was driven by diesel engine on the surface and by battery-powered electric motors underwater. In 1954 the USA launched the first nuclear-powered submarine, the *Nautilus*.

See also: *midget submarine; U-boat; X-craft.*

NUCLEAR SUBMARINE

- The US nuclear submarine Ohio, in service from 1981, is 170 m/560 ft long and carries 24 Trident missiles, each with 12 independently targetable nuclear warheads.
- Three Vanguard-class, Trident missile-carrying submarines, which when armed will each wield more firepower than was used in the whole of World War II, are being built in the UK.

Suvla Bay

Bay in **Gallipoli**, west of the Dardanelles, scene of fierce fighting between Turkish and British and Commonwealth troops during **World War I**.

Four British divisions were landed there on 6 August 1915 to capture the Anafarta Hills and give the Allies control of the central heights of the peninsula. However, by the time orders were given to attack, the Turks had been strongly reinforced with more troops and artillery and the attack failed. Further attacks were mounted over the next few days, but no impression could be made on the Turkish positions and the Allied lines settled down to defend what they had.

sword

Sharp-edged weapon used in hand-to-hand fighting for cutting or stabbing. It consists of a blade and a hilt, or handle.

The first swords were made in bronze, but by 1,000 BC sword-makers were using iron. In the Middle Ages swords were being made in Europe and Japan up to 1.8m/6 ft long for wielding with one hand or both hands. The use of swords declined after the invention of firearms, but the sabre, a sword with a curved blade, continued to be used by cavalry into the 1900s.

CLAYMORES, RAPIERS, AND SCIMITARS

- Swords vary greatly from one another: from the broad-bladed, double-edged Scottish claymore to the narrow, sharp-pointed rapier.
- The Persian scimitar has a highly curved blade that widens towards the point.

Talavera, Battle of
British and Spanish victory over the French near Talavera de la Reina, a Spanish town 110 km/70 mi southwest of Madrid, on 27–28 July 1809, during the **Peninsular War**. The Duke of **Wellington** drove off the French force under Marshal Nicolas Soult, but the British were so weakened by the battle they were unable to pursue Soult, and Wellington fell back into Portugal. The British lost some 6,200 killed and wounded; the French about 7,400 killed and wounded.

tank
Armoured fighting vehicle that runs on caterpillar tracks to enable it to cross rough ground, and is fitted with weapons systems capable of defeating other tanks and destroying life and property. It was invented by the British soldier and scholar Ernest Swinton and first used in the Battle of the **Somme** in 1916.

WATER CARRIER

- A tank consists of a body or hull of thick steel, on which are mounted machine guns and a larger gun. The hull contains the crew (usually consisting of a commander, driver, and one or two soldiers), engine, radio, fuel tanks, and ammunition.
- The name arose from the cover used when developing the British prototype. In an endeavour to keep the project secret, the test model was described as a 'water carrier for Mesopotamia', from which it became known in the factory as 'that tank thing'.

Tannenberg, Battle of
Victory, in 1410, of a combined Polish and Lithuanian army over the Knights of the Teutonic Order at Tannenberg, a village in northern Poland (now Grunwald). King Wladyslaw Jagiello, Grand Duke of Lithuania and

king of Poland, led an army of 20,000 to meet some 15,000 Knights and routed them, slaughtering several thousand. The defeat broke the Knights' hold over Old Prussia (approximately modern Poland) and led to the Treaty of Thorn and to an independent Polish state.

Tannenberg, Battle of
World War I victory of German forces led by field marshal Paul von Hindenburg over Russian forces under Gen Aleksander Samsonov in August 1914 at a village in East Prussia (now Grunwald, Poland) 145 km/90 mi northeast of Warsaw.

Hindenburg surrounded Samsonov on three sides at Tannenberg, the fourth being swamps and lakes, and destroyed his army; only about 60,000 troops managed to escape back to Russia. The Germans took 90,000 prisoners and several hundred guns.

Tet Offensive
In the **Vietnam War**, a prolonged attack mounted by the Vietcong against Saigon (now Ho Chi Minh City) and other South Vietnamese cities and hamlets (including the US Marine base at Khe Sanh), which began on 30 January 1968. Although the Vietcong were finally forced to withdraw, the Tet Offensive brought into question the ability of the South Vietnamese army and their US allies to win the war and added fuel to the antiwar movement in both the USA and Australia. In this regard, the Tet Offensive might be considered the watershed of the Vietnam War.

> Of 84,000 communist Vietcong who took part in the offensive, 32,000 were killed by mid-February.

Thirty Years' War
A major war in central Europe in 1618–48. Beginning as a conflict between Protestants and Catholics in Bohemia, it was gradually transformed into a struggle to determine whether Ferdinand II, the Habsburg monarch and later Holy Roman Emperor, could gain control of all Germany. The German Protestants, who received support from England, Denmark, and the United Provinces, were saved by the intervention of Sweden, and in 1635 France entered the conflict to curb the power of the Empire. Under the Peace of Westphalia, in 1648, the German states were granted their sovereignty and the Emperor retained only in nominal control. The war caused serious economic and demographic problems in central Europe.

Main events of the Thirty Year's War.

1618 Protestant nobles depose Catholic King Ferdinand in Prague.
1620–23 Protestants are defeated by Imperialist forces.
1625 Invasion of Germany by Denmark in support of the Protestants.
1629 Denmark concludes peace after serious defeats.
1631–34 Sweden invades and defeats Imperialist forces at Breitenfeld (1631) and Lützen (1632), but is beaten at Nordlingen (1634).
1635 Peace made at Prague. Intervention of France restarts the war.
1640–48 Peace of Westphalia ends war, after major French victories.

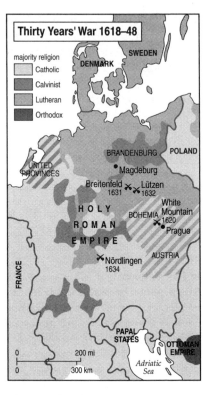

Tobruk, Battles of

A series of engagements in **World War II** between British and **Axis** forces in the struggle for control of the Libyan port of Tobruk.

Occupied by Italy since 1911, Tobruk was taken by Britain in Operation Battleaxe in 1941, and unsuccessfully besieged by Axis forces in April–December 1941. It was captured by Germany in June 1942 after the retreat of the main British force to Egypt, and this precipitated Gen Sir Claude Auchinleck's replacement by Gen Bernard **Montgomery** as British commander. Montgomery recovered Tobruk after the second Battle of El **Alamein** and it remained in British hands for the rest of the war.

torpedo

Self-propelled underwater **missile**, invented by British engineer Robert Whitehead in 1866. Modern torpedoes are homing missiles; some resemble mines in that they lie on the seabed until activated by the acoustic signal of a passing ship. A television camera enables them to be remotely controlled, and in the final stage of attack they lock onto the radar or sonar signals of the target ship.

Tours, Battle of or Battle of Poitiers

A battle fought near Poitiers, France, in 732, between the Franks, under Charles Martel, and the Muslims. The outcome was a decisive victory for the Franks. Although the Muslims again raided into southern France, the Battle of Tours (or more accurately the Battle of Poitiers) ended the possibility of further Muslim conquests in France and Western Europe.

> **MUSLIM DEFEAT**
>
> - The Franks dismounted their cavalry and formed a solid shield wall, which successfully beat off successive Muslim cavalry charges.
> - The Muslims began to withdraw, but when their leader Abd ar-Rahman was killed in the fighting, the withdrawal turned into a rout.

Trafalgar, Battle of

A victory of the British fleet, commanded by Admiral Horatio **Nelson**, over a combined French and Spanish fleet, under Admiral Pierre de Villeneuve, on 21 October 1805, during the **Napoleonic Wars**. Nelson was mortally wounded during the action. The battle is named after Cape Trafalgar, a low headland in southwest Spain, near the western entrance to the Straits of Gibraltar.

The French were sailing in a loose-line formation and Nelson divided his force into two parts, which he drove through the French line at different points. The manoeuvre was successful, Nelson's flagship *Victory* passing the stern of the French flagship *Bucentaure* and discharging its broadside at a range of 11 m/12 yd, causing 400 casualties. Other British ships used similar tactics of close-quarter gunnery. The battle commenced at about 12 noon, and by 3 pm it was over.

> **NO BRITISH SHIPS LOST**
>
> - Of the 33 French and Spanish ships, 15 were sunk, 2 were wrecked on 24 October, and 4 taken by a British squadron on 3 November. French and Spanish casualties amounted to about 14,000.
> - The British lost none of their 27 ships but sustained casualties of 449 killed and 1,242 wounded.

> England has saved herself by her exertions, and will, as I trust, save Europe by her example.

William Pitt the Younger, British Tory prime minister, on the Battle of Trafalgar.

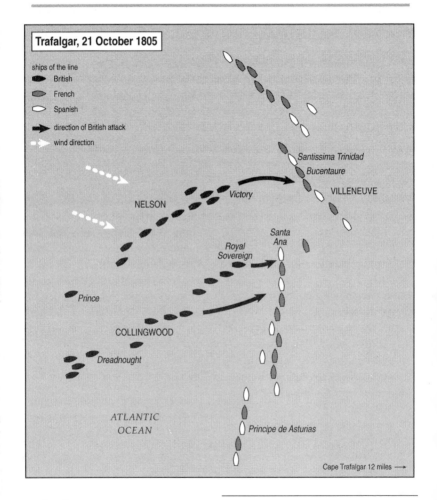

trebuchet

A missile-launching weapon resembling a catapult, with a

The most powerful trebuchets could hurl missiles weighing a tonne or more.

beam pivoted on an axle and finishing in a sling. It was invented in China between the 5th and 3rd centuries BC, and reached Europe around AD 500.

Trenchard, Hugh Montague, 1st Viscount Trenchard (1873–1956) British aviator and police commissioner. In 1915–17, during **World War I,** he commanded the Royal Flying Corps, and from 1918 to 1929 he organized the Royal Air Force, becoming its first marshal in 1927.

As commissioner of the Metropolitan Police in 1931–35, he established the Police College at Hendon and carried out the Trenchard Reforms, which introduced more scientific methods of detection. He was knighted in 1918, and made Viscount in 1936.

Trident

A nuclear missile deployed on US nuclear-powered submarines, and in the 1990s installed on four UK submarines. Each missile has eight warheads **(MIRVs).** The four UK submarines will each carry 16 Trident D-5 missiles. The Trident replaced the earlier Polaris and Poseidon missiles.

> The first British submarine designed to carry Tridents was launched 1992 and is of the giant Vanguard class.

Tsushima, Battle of

Japanese naval victory over the Russians on 27–28 May 1905, during the **Russo–Japanese War**. The battle was fought in the Strait of Tsushima between Japan and Korea.

The Russian fleet, consisting of 46 ships, of which 7 were Dreadnought-type battleships and 6 cruisers, was met by a Japanese fleet of similar size and composition but capable of greater speed and with better-trained sailors. In the battle, all but 12 minor ships of the Russian fleet were sunk, captured, or driven aground, while the Japanese lost only three torpedo-boats.

> The Battle of Tsushima was the only engagement between Dreadnought-type battleships, and arguably one of the greatest naval battles ever fought.

Turenne, Henri de la Tour d'Auvergne, Vicomte de Turenne (1611–1675)
French marshal under Louis XIV, renowned for his siege technique. He fought for the Protestant alliance during the **Thirty Years' War**, and on both sides during the civil war of 1648–52 known as the Fronde.

U-2
US military reconnaissance aeroplane. From 1956, during the Cold War, it was used in secret flights over the USSR to photograph military installations. In 1960 a U-2 was shot down

Designed by Richard Bissell, the U-2 flew higher (21,000 m/70,000 ft) and further (3,500 km/2,200 mi) than any previous plane.

over the USSR and the pilot, Gary Powers, was captured and imprisoned. He was exchanged for a US-held Soviet agent two years later. In 1962, U-2 flights revealed the construction of Soviet missile bases in Cuba.

U-boat
German **submarine** named from the German *Unterseeboot*/Undersea boat. The term was used in both **World Wars I** and **II**, when U-boat attacks posed a great threat to Allied shipping.

Unknown Soldier
Unidentified dead soldier, for whom a tomb is erected as a memorial to other unidentified soldiers killed in war.

In Britain, the practice began in **World War I**; the British Unknown Soldier was buried in Westminster Abbey 1920. France, Belgium, the USA, and other countries each have Unknown Soldier tombs.

V1, V2

German flying bombs of **World War II**, launched against Britain in 1944 and 1945 and named after the German *Vergeltungswaffe* 'revenge weapons'. The V1, also called the doodlebug and buzz bomb, was an uncrewed monoplane carrying a bomb, powered by a simple kind of jet engine called a pulse jet. The V2, a rocket bomb with a preset guidance system, was the first long-range ballistic missile.

VON BRAUN AND THE SPACE RACE

- The V2 was developed by the rocket engineer Wernher von Braun (1912–1977).
- It was 14 m/47 ft long, carried a 1-tonne warhead, and hit its target at a speed of 5,000 kph/3,000 mph.
- After World War II, captured V2 material became the basis of the space race in both the USSR and the USA.

Vauban, Sébastien le Prestre de (1633–1707)

French marshal and military engineer. In Louis XIV's wars he engineered many successful **sieges** and rebuilt many of the fortresses on France's east frontier. He also built new towns at key defensive points, such Neuf-Brisach, near the German border east of Colmar.

V bombs

See **V1, V2**.

Verdun

Fortress town in northeast France in the *département* of the Meuse, 280 km/174 mi east of Paris. During **World War I** it became a symbol of French resistance and was the centre of a series of bitterly fought actions between French and German forces, finally being recaptured September 1918.

In 1916 the Germans attacked it in great strength. The battle continued for the rest of the year, both sides capturing and recapturing forts and ground. The French lost an estimated 348,300 troops, the Germans 328,500. The French regained the Mort Homme area by 12 September 1917, but it was not until September 1918 that a decisive attack was launched as part of the general Allied offensive.

Vicksburg, Battle of

In the American Civil War, a Union victory over Confederate forces May–July 1863, at Vicksburg, Mississippi, 380 km/235 mi north of New Orleans. Vicksburg, on the Mississippi River, was a well-fortified communications hub of great importance and the Union capture of the town virtually split the Confederacy in two. It also brought Ulysses S **Grant** to public prominence, eventually leading to him being given command of all Union forces.

UNDER SIEGE

- Two Union corps under Gen John A McClernand laid siege to Vicksburg in January 1863.
- In May, McClernand was relieved by Grant, who took personal command of the siege.
- Grant pressed the town hard until Lt-Gen John C Pemberton, the Confederate commander, surrendered on 4 July with 25,000 troops and 90 guns.

Vienna, Battle of

An unsuccessful siege of Vienna in September–October 1529 by the Turks, commanded by Suleiman the Magnificent. Vienna marked the farthest extent of the Ottoman invasion of the West.

Vienna was held by a garrison of about 16,000 soldiers, when an army of 120,000 Turks besieged it. Several desperate assaults were made and repulsed, and Turkish artillery bombarded the walls and eventually breached them. A final attempt by the Turks to storm this breach and enter the city was beaten off with heavy casualties on the Turkish side. Suleiman consequently abandoned the siege and retired east.

Vietnam War

A war from 1954 to 1975 between North Vietnam and US-backed South Vietnam, in which North Vietnam aimed to unite the country as a communist state.

Consequently Noncommunist South Vietnam was viewed, in the context of the 1950s and the **Cold War**, as a bulwark against the spread of communism throughout Southeast Asia.

The USA spent $141 billion on aid to the South Vietnamese government, but corruption and inefficiency led the USA to assume ever greater responsibility for the war effort, until 1 million US combat troops were engaged. At the end of the war, the USA withdrew, and North and South Vietnam were reunited as a socialist republic. Although US forces were never militarily defeated, Vietnam was considered a most humiliating political defeat for the USA.

DEATH AND DESTRUCTION

- Some 200,000 South Vietnamese soldiers, 1 million North Vietnamese soldiers, and 500,000 civilians were killed.
- In 1961 to 1975, 56,555 US soldiers were killed, a fifth of them by their own troops.
- Cambodia, a neutral neighbour of Vietnam, was bombed by the USA in 1969–75, with 1 million killed or wounded.
- The war destroyed 50% of Vietnam's forest cover and 20% of its agricultural land.

Vimiero, Battle of

A defeat of the French by the Duke of **Wellington** during the **Peninsular War**, on 21 August 1808, near Vimiero, a Portuguese village 50 km/31 mi northwest of Lisbon. Wellington held a strong position, and although the French attack was pushed hard, it was beaten off and the French fell back in disorder. Wellington wanted to pursue the defeated French, but his superior Sir Henry Burrard refused permission. They escaped and were allowed to leave Portugal in English ships under the terms of the Convention of Cintra, allowing the Peninsular War to drag on for several more years.

Vimy Ridge

A hill in northern France, taken in **World War I** by Canadian troops during the Battle of Arras, in April 1917, at the cost of 11,285 lives. It is a spur of the ridge of Notre Dame de Lorette, 8 km/5 mi northeast of Arras.

Vitoria, Battle of

A French defeat by the **Duke of Wellington** during the **Peninsular War**, on 21 June 1813, near Vitoria, a north Spanish town. This battle effectively ended French influence in Spain, and after clearing out French stragglers from the border area, the British were able to march on into France

Wellington had re-entered Spain, having wintered in Portugal, and marched across northern Spain. With a combined British–Portuguese–Spanish force of about 75,000 troops and 90 guns, he outflanked the French. After a hard fight, the French retired toward Pamplona, having lost 8,000 casualties, 2,000 prisoners, and 152 guns; Wellington lost 4,500 killed and wounded.

Wagram, Battle of
Major victory by **Napoleon**, on 5–6 July 1809, during the Napoleonic Wars, over an Austrian army led by Archduke Charles of Austria near Wagram, northeast of Vienna. The battle was famous for the heavy concentration of field artillery, the largest-ever seen up to that time.

Wake Island (also called Enenkio)
A small Pacific coral atoll, 8 sq km/3 sq mi in area, comprising three islands 3,700 km/2,300 mi west of Hawaii, under US Air Force administration since 1972. It was discovered by Captain William Wake in 1841, annexed by the USA in 1898, and uninhabited until 1935, when it was made an air staging point, with a garrison. It was occupied by Japan in 1941–45.

Wallenstein, Albrecht Eusebius Wenzel von (1583–1634)
German general who, until his defeat at Lützen in 1632, led the Habsburg armies in the **Thirty Years' War**. Two years later, he was assassinated.

war
Act of force, usually on behalf of the state, intended to compel a declared enemy to obey the will of the other. The aim is to render the opposition incapable of further resistance by destroying its capability and will to bear arms in pursuit of its own aims. War is therefore a continuation of politics carried on with violent and destructive means, as an instrument of policy. *See also: **civil war, guerrilla warfare**.*

Other types of war include:

- Limited war a war limited in both geographical extent and levels of force used whose aims stop short of achieving the destruction of the enemy. The Korean War falls within this category.

❦ After each war there is a little less democracy to save. ❧

Brooks Atkinson, US journalist, on war in *Once Around the Sun*

- *Total war* the waging of war against both combatants and non-combatants. The Spanish Civil War, in which bombing from the air included both civilian and military targets, marked the beginning of this type of warfare.
- *Absolute war* a concept of war that allows no limitations, such as law, compassion, or prudence, in the application of force, the sole aim being to achieve the complete annihilation of the enemy.
- *Major armed conflict* a conflict that kills more than 1,000 people a year.

CIVILIAN VICTIMS

- In the wars of the late 20th century, 90% of casualties have been civilian (in World War II, the figure was 50%; in World War I only 5%).
- The estimated figure for loss of life in wars in the Third World since 1945 is 17 million.

❝When war is declared, Truth is the first casualty.❞

Arthur Ponsonby, British politician and author, on war in *Falsehood in Wartime*, epigraph.

war crime

Offence (such as murder of a civilian or a prisoner of war) that contravenes the internationally accepted laws governing the conduct of wars, particularly the Hague Convention of 1907 and the **Geneva Convention** of 1949. A key principle of the law relating to such crimes is that obedience to the orders of a superior is no defence. In practice, prosecutions are generally brought by the victorious side.

In 1943 the United Nations War Crimes Commission was set up to investigate atrocities against Allied nationals in **World War II**. Leading Nazis were defendants in the Nuremberg

In November 1995 the Bosnian Serb leader Radovan Karadzic and his general Ratko Mladic were charged, in their absence, with **genocide** and crimes against humanity at the **Yugoslav War** Crimes Tribunal in The Hague.

trials of 1945–46. High-ranking Japanese officers were tried in Tokyo before the International Military Tribunal. The hunt for Nazis who escaped justice has continued, led notably by Simon Wiesenthal (1909–), who tracked down Adolf Eichmann in 1960.

war graves
Graves of soldiers who fell during **World War I**, buried in the war zones. Vast cemeteries were established after the war by the countries that fought in the various theatres, and in France and Belgium, where most war graves lie, the land was presented to the countries whose men are buried there. Each nation established its own Graves Commission to supervise the construction, interment, and maintenance of the cemeteries, which has continued to the present day. It is noteworthy that, without formal agreement, the warring armies avoided damaging these cemeteries during **World War II**.

War of 1812
The war between the USA and Britain caused by British interference with US merchant shipping as part of Britain's economic warfare against **Napoleonic** France. Tensions between the Americans and the British in Canada led to plans for a US invasion but these were never realized. In 1814 British forces occupied Washington, DC, and burned the White House and the Capitol. A treaty signed in Ghent, Belgium, in December 1814 ended the conflict.

> Before news of the peace treaty signed in Ghent reached the USA, American troops under Andrew Jackson defeated the British at New Orleans in 1815.

Wars of the Roses
See **Roses, Wars of the**.

Warsaw rising
The uprising of August–October 1944 against German occupation of Warsaw in **World War II**. The rebellion, which was organized by the Polish Home Army, was brutally quashed when anticipated Soviet help for the rebels did not arrive.

The German army had begun withdrawing from Warsaw in anticipation of the arrival of the Soviets when the Home Army rose to keep the German troops occupied to make it easier for the Soviets to enter the city. Street fighting began 1 August, but on the following day the Soviet attack was halted and the Germans were free to turn their full power against the rebellion.

warship

Fighting ship armed and crewed for war. The supremacy of the **battleship** at the beginning of the 20th century was rivalled during **World War I** by the development of **submarine** attack, and was rendered obsolete in **World War II** with the advent of long-range air attack. Today the largest and most important surface warships are the **aircraft carriers**.

The modern warship carries three types of sensor:

- radar for surface-search and tracking, navigation, air surveillance, and indication of targets to weapon-control systems
- sonar for detection of surface and subsurface targets
- liod (lightweight optronic detector) for processing the optical contrast of a target against its background, viewed by a television or infrared camera.

See also: *destroyer; frigate.*

Washington, George (1732–1799)

Commander of the American forces during the **American Revolution** and 1st president of the USA in 1789–97; known as 'the father of his country'.

An experienced soldier, he had fought in campaigns against the French during the French and Indian War. He was elected to the Virginia House of Burgesses in 1759 and was a leader of the Virginia militia, gaining valuable experience in wilderness fighting. As a strong opponent of the British government's policy, he sat in the Continental Congresses of 1774 and 1775, and on the outbreak of

Washington, George *Portrait of Washington*

the American Revolution was chosen as commander-in-chief of the Continental army. After many setbacks, he accepted the surrender of British general Cornwallis at **Yorktown** in 1781.

Elected president in 1789, he was re-elected in 1793, but declined a third term. In his farewell address, he warned the country against entangling alliances and to remain aloof from European quarrels.

> On 14 December 1999, exactly 200 years after Washington's death at his home in Mount Vernon, his last hours were reenacted in a ceremony in his bedroom, by descendants of Washington, his wife, and her slave. Four days later, the public was invited to join dignitaries at Mount Vernon to dine on pound cake, as mourners did in 1899.

Waterloo, Battle of

Final battle of the **Napoleonic Wars**, in which, on 18 June 1815, a coalition force of British, Prussian, and Dutch troops under the Duke of **Wellington** defeated **Napoleon** near the village of Waterloo, 13 km/8 mi south of Brussels, Belgium. Napoleon found Wellington's army isolated from his allies and began a direct offensive to smash them, but the British held on until joined by the Prussians under Marshal Gebhard von Blücher.

Wellington had 67,000 soldiers (of whom 24,000 were British, the remainder being German, Dutch, and Belgian) at Waterloo, and Napoleon had 74,000. The French casualties numbered about 37,000; coalition casualties were similar including some 13,000 British troops.

> ❝ I have got an infamous army, very weak and ill-equipped, and a very inexperienced staff. ❞
>
> **Arthur Wellesley**, 1st Duke of Wellington, Letter 8 May 1815, just before the battle of Waterloo.

weapon

Implement or device used for doing damage to an enemy. Down the ages, weapons have ranged from simple clubs, **spears**, and **bows and arrows** to **cannon**, **machine guns**, and nuclear **missiles**.

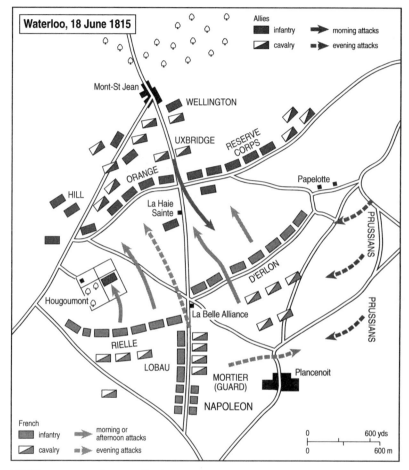

Wellington, Arthur Wellesley, 1st Duke of Wellington (1769–1852)

Irish-born British soldier, Tory prime minister, and victor at the Battle of Waterloo. In 1804, he was knighted for his distinguished army service in India. As commander in the Peninsular War, he expelled the French from Spain,

NEXT TO NELSON

- Born in Dublin, Wellington served in the early 1790s in the Irish parliament as the member for Trim.
- He is buried in St Paul's Cathedral by the side of Nelson.

became a national hero, and in 1814 was made Duke. He defeated Napoleon Bonaparte at Quatre-Bras and Waterloo in 1815, and was a member of the Congress of Vienna. Prime minister in 1828–30, he modified the Corn Laws but was forced to concede Roman Catholic emancipation.

Wellington, *Arthur Wellesly.*

❝ I always say that, next to a battle lost, the greatest misery is a battle gained. ❞

Arthur Wellesley, 1st Duke of Wellington, attributed remark.

Wilderness, Battle of the
An indecisive battle between Union and Confederate forces on 4–8 May 1864 in a wooded area known as 'The Wilderness', about 24 km/15 mi west of Fredericksburg, Virginia, during the American **Civil War**.

At the start of the 1864 campaigning season, the Confederate general Robert E **Lee**, holding a position behind the Rapidan River in the Wilderness area, was confronted by Gen Ulysses S **Grant** who had been given command of the Union armies. The battle ended with Grant slipping his army sideways to cross the Rapidan near Spotsylvania where Lee anticipated him and a new battle began.

Wingate, Orde Charles (1903–1944)
British soldier who fought in Burma (now Myanmar) in **World War II**. In 1936–39 he organized Jewish **guerrillas** in Palestine. In World War II he served in the Middle East and led guerrilla forces in Ethiopia, and later led

the Chindits, the Third Indian Division, in guerrilla operations against the Japanese army in Burma. Wingate's airborne troops caused havoc behind the Japanese lines, disrupting Japanese communications for three months from March 1944. He died in an air crash.

Wolfe, James (1727–1759)
English soldier who served in Canada and commanded a victorious expedition against the French general Montcalm in Quèbec on the Plains of Abraham. During the battle both commanders were killed. The British victory established British supremacy over Canada.

Wolfe fought at the battles of Dettingen, Falkirk, and **Culloden**. With the outbreak of the **Seven Years' War**, he was posted to Canada and played a conspicuous part in the siege of the French stronghold of Louisburg in 1758. He was promoted to major general in 1759.

Women's Land Army
In Britain, an organization founded in 1916 for the recruitment of women to work on farms during **World War I**. At its peak in September 1918 it had 16,000 members. It re-formed in June in 1939, before the outbreak of **World War II**. Many 'Land Girls' joined up to help the war effort and, by August 1943, 87,000 were employed in farm work.

women's services
The organized military use of women on a large scale, a 20th-century development. First, women replaced men in factories, on farms, and in non-combat tasks during wartime. They are now found in combat units in many countries, including the USA, Cuba, the UK, and Israel.

In Britain there are women's corps for all three armed services:
Women's Royal Army Corps (WRAC) created 1949 to take over the functions of the Auxiliary Territorial Service, established 1938; its **World War I** equivalent was the Women's Army Auxiliary Corps (WAAC).
Women's Royal Naval Service (WRNS) active in 1917–19 and 1939 onwards, allowed in combat roles on surface ships from 1990.
Women's Royal Air Force (WRAF) established in 1918 but known in 1939–48 as the Women's Auxiliary Air Force (WAAF).

World War I
War in 1914–18, also known as the Great War, between the Central European Powers (Germany, Austria-Hungary, and allies) on one side and

the Triple Entente (Britain and the British **Empire**, France, and Russia) and their allies, including the USA (which entered 1917), on the other side. It was fought on the eastern and western fronts, in the Middle East, in Africa, and at sea.

An estimated 10 million lives were lost in World War I and twice as many people were wounded.

By the early 20th century, the countries of Western Europe had reached a high level of material prosperity. However, competition for trade markets and imperial possessions worldwide led to a growth of nationalistic sentiment. This nationalism created great political tension between the single-nation states such as France and Germany, and threatened the stability of multi-nation states such as Austria–Hungary.

The war was set in motion by the assassination in Sarajevo of the heir to the Austrian throne, Archduke Franz Ferdinand, by a Serbian nationalist in June 1914. When Austria declared war on Serbia on 28 July 1914, Russia mobilized along the German and Austrian frontier. Germany then declared war on Russia and France. On 4 August, Britain declared war on Germany.

An armistice was signed between Germany and the Allies at 5 am on 11 November 1918, and fighting ceased on the Western Front at 11 am the same day.

World War I: Chronology

1914 *June*: Assassination of Archduke Franz Ferdinand of Austria, 28 June.

July: Germany offers Austria, offering support in war against Serbia. Russia begins mobilization to defend Serbian ally.

August: Germany declares war on Russia. France mobilizes to assist Russian ally. Germany declares war on France and invades Belgium. Britain declares war on Germany, then on Austria. Dominions within the British Empire, including Australia, are automatically involved.

September: British and French troops halt German advance just short of Paris, and drive them back. First Battle of the Marne, and of the Aisne. Beginning of trench warfare.

October–November: First Battle of Ypres. Britain declares war on Turkey.

1915 *April–May*: Gallipoli offensive launched by British and dominion troops against Turkish forces. Second Battle of Ypres. First use of poison gas by Germans. Italy joins war against Austria.

1916 *February*: German offensive against Verdun begins, with huge losses for small territorial gain.

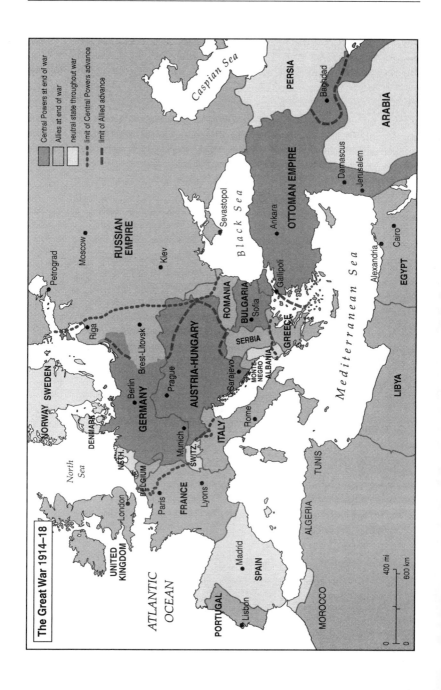

	May: Naval Battle of Jutland between British and German imperial fleets ends inconclusively.
	July–November: First Battle of the Somme, a sustained Anglo-French offensive which wins little territory and costs a huge number of lives.
	September: Early tanks are used by British on Western Front.
1917	*February*: Germany declares unrestricted submarine warfare. Russian Revolution begins and tsarist rule is overthrown.
	March–April: Germans retreat to Siegfried Line (Arras-Soissons) on Western Front.
	April–May: USA enters the war against Germany.
	July–November: Third Ypres offensive including Battle of Passchendaele.
	September: Germans occupy Riga.
	December: Jerusalem taken by British forces under Allenby.
1918	*January*: US President Woodrow Wilson proclaims 'Fourteen Points' as a basis for peace settlement.
	March: Treaty of Brest-Litovsk with Central Powers ends Russian participation in the war. Second Battle of the Somme begins with German spring offensive.
	July–August: Allied counter-offensive, including tank attack at Amiens, drives Germans back to the Siegfried Line.
	September: Germany calls for an armistice.
	October: Armistice offered on the basis of the 'Fourteen Points'.
	November: Austria-Hungary signs armistice with Allies. Germany agrees armistice. Fighting on Western Front stops.
1919	*January*: Peace conference opens at Versailles.
	May: Demands are presented to Germany.
	June: Germany signs peace treaty at Versailles, followed by other Central Powers.

> ❝ Would you kindly forward the enclosed letter and earn the blessing of a poor British soldier? ❞
>
> Note enclosed with a letter that was 'posted' in a bottle by **Pte Thomas Hughes**, 26, to his wife in 1914, 12 days before he was killed. It was found in the Thames Estuary in April 1999 by fisherman Steve Gowan, and hand delivered by him to Pte Hughes's daughter in Auckland, New Zealand, on 17 May 1999; *Daily Telegraph*, 18 May 1999.

World War II

War 1939–45 between Germany, Italy, and Japan (the Axis powers) on one side, and Britain, the Commonwealth, France, the USA, the USSR, and China (the Allied powers) on the other. The war was fought in the Atlantic and

Pacific theatres. In May 1945, Germany surrendered but Japan fought on until August, when the USA dropped **atomic bombs** on **Hiroshima** and Nagasaki.

Under Adolf **Hitler**, Germany embarked on a programme of aggressive expansionism. Britain and France declared war on Germany on 3 September 1939, two days after German forces had invaded Poland. In the following months (the 'phoney' war) little fighting took place until April 1940, when the Germans invaded Denmark and Norway. By the end of May, Germany had invaded Holland, Belgium, and France, and 337,131 Allied troops had to be evacuated from **Dunkirk** to England. Following the German aerial bombardment of British cities known as the **Blitz**, the RAF averted a planned invasion of Britain in the **Battle of Britain**.

The major turning point for the Allies was victory in the Battle of El **Alamein** in 1942. The Allies launched the successful D-Day invasion of Normandy on 6 June 1944 under the command of US general Eisenhower. By spring 1945, the Allied advances from the west and east had joined. All German forces in northwest Germany, Holland, and Denmark surrendered to Field Marshal **Montgomery** on 5 May, and Germany's final capitulation came into effect at midnight on 8–9 May.

DEAD AND DISPLACED

- An estimated 55 million lives were lost in World War II, including 20 million citizens of the USSR and 6 million Jews killed in the Holocaust.
- Some 60 million people in Europe were displaced because of bombing raids.

World War II: Chronology

1939 *September*: German invasion of Poland; Britain and France declare war on Germany.
1940 *April*: Germany occupies Denmark, Norway, the Netherlands, Belgium, and Luxembourg.
May–June: Evacuation of Allied troops from Dunkirk to England.
June: Italy declares war on Britain and France; the Germans enter Paris.
July–October: Battle of Britain between British and German air forces.
September: Japanese invasion of French Indochina.
1941 *April*: Germany occupies Greece and Yugoslavia.
June: Germany invades the USSR; Finland declares war on the USSR.
December: The Germans come within 40 km/25 mi of Moscow, with Leningrad (now St Petersburg) under siege. Japan bombs Pearl Harbor, Hawaii, and declares war on

the USA and Britain. Germany and Italy declare war on the USA.

1942 *January*: Japanese conquest of the Philippines.

June: Naval battle of Midway, the turning point of the Pacific War.

August: German attack on Stalingrad (now Volgograd), USSR.

October–November: Battle of El Alamein in North Africa.

November: Soviet counteroffensive on Stalingrad.

1943 *May*: End of Axis resistance in North Africa.

August: Beginning of the campaign against the Japanese in Burma (now Myanmar).

September: Italy surrenders to the Allies.

October: Italy declares war on Germany.

November: The US Navy defeats the Japanese in the Battle of Guadalcanal.

November–December: The Allied leaders meet at the Tehran Conference.

1944 *January*: Allied landing in Nazi-occupied Italy: Battle of Anzio.

March: End of the German U-boat campaign in the Atlantic.

6 June: D-day: Allied landings in Nazi-occupied and heavily defended Normandy.

September: Battle of Arnhem on the Rhine; Soviet armistice with Finland.

December: German counteroffensive, Battle of the Bulge.

1945 *February:* The Soviets reach the German border; Yalta conference; Allied bombing campaign over Germany (Dresden destroyed); the US reconquest of the Philippines is completed; the Americans land on Iwo Jima, south of Japan.

April: Hitler commits suicide; Mussolini is captured by Italian partisans and shot.

May: Germany surrenders to the Allies.

July: The Potsdam Conference issues an Allied ultimatum to Japan.

August: Atom bombs are dropped by the USA on Hiroshima and Nagasaki; Japan surrenders.

❛ Overpaid, overfed, oversexed, and over here. ❜

Tommy Trinder, English entertainer on US troops in Britain during World War II, in *The Sunday Times*, 4 January 1976.

WORLD WAR II · 197

X-craft

British midget submarines of **World War II**. Introduced in 1943, they were intended to enter restricted harbours and lay time-fused explosive charges beneath enemy ships.

Displacing about 30 tons-30.5 tonnes, they had a crew of four or five and were propelled by electric motors giving a speed of 6 knots (11 kph). They were used to immobilize the *Tirpitz* in September 1943 and also against Japanese ships in **Singapore** harbour in July 1945.

Yalu River, Battle of
In the first **Sino-Japanese War**, Chinese naval defeat by a Japanese fleet on 17 September 1894 at the mouth of the Yalu River, the border between Korea and Manchuria. This was the first large-scale naval action in which breech-loading guns, quick-firing guns, and torpedoes were used.

Yom Kippur War
The Fourth Arab–Israeli War, which began in October 1973 when Egypt and Syria launched a surprise attack on Israel on Yom Kippur, a Jewish national holiday and the holiest day of the Jewish year. *See* **Arab–Israeli Conflict**.

Yorktown, Battle of
Decisive defeat for the British at Yorktown, Virginia, 105 km/65 mi southeast of Richmond, in September–October 1781 during the **American Revolution**. The British commander Lord Cornwallis had withdrawn into Yorktown where he was besieged by 7,000 French and 8,850 American troops and could only wait for reinforcements to arrive by sea. However, the Royal Navy lost command of the sea at the Battle of **Chesapeake** and with no reinforcements or supplies forthcoming, Cornwallis was forced to surrender on 19 October, effectively conceding victory to the Americans in the war.

Ypres, Battles of
Three major **World War I** battles 1914–1917 between German and Allied forces near Ypres, a Belgian town in western Flanders, 40 km/25 mi south of Ostend. Neither side made much progress in any of the battles, despite heavy casualties. The third battle in particular (*see* **Passchendaele**) stands out as an enormous waste of life for little return. The Menin Gate (1927) is a memorial to British soldiers lost in these battles.

- *October–November 1914* A British offensive aimed at securing the Channel ports of Dunkirk and Ostend clashed with a German offensive with a similar aim.

- *April–May 1915* A German attack with chlorine gas made a gap in the Allied lines but the Allies rushed in reserves.
- *July–November 1917* An Allied offensive under Field Marshal Douglas Haig to capture ports on the Belgian coast held by Germans gained only 8 km/5 mi of territory at the price of more than 300,000 casualties.

Yugoslav War

A series of bitter civil wars precipitated by nationalist or ethnic moves for independence or autonomy that attended the break-up of federal Yugoslavia after the death, in 1980, of the long-serving Communist president Tito.

- *Slovenia* Fighting broke out between republican and federal forces after the Slovenian republic declared independence from Yugoslavia in 1991. In 1992 the Yugoslav president, Milosevic, withdrew his forces under economic pressure from the European Union and Slovenia's independence was recognized.
- *Croatia* A Serb minority, who wanted Croatia to remain part of Yugoslavia and was backed by the federal army, fought against the majority Croats who, headed by president Tudjman, wanted independence.
- *Bosnia-Herzegovina* In 1992, pro-independence Bosnian Muslims and Croats opposed pro-federal Bosnian Serbs fighting to partition the newly created state. The latter were accused of **war crimes** and 'ethnic cleansing' in the conflict. Yugoslavia finally recognized Bosnia-Herzegovina's and Croatia's independence in 1995.
- *Kosovo* Violence between the majority (90%) Muslim Albanians in Kosovo (a province of Yugoslavia) and minority Serbs led to intervention by the Yugoslav army against the KLA (Kosovo Liberation Army), and to ethnic cleansing of Albanians. In 1999, after stalled peace talks involving the United Nations, NATO aircraft bombed Kosovo. Hundreds of thousands of ethnic Albanians fled from the Serbs and the bombing into neighbouring countries, creating a refugee crisis. In June 1999, President Milosevic accepted NATO's peace terms, under which NATO forces took control of Kosovo to keep the peace between the two communities.

Zama, Battle of
Battle fought in 202 BC between the Roman army of Publius Cornelius Scipio 'Africanus' and the Carthaginian army of **Hannibal** the Great, at Zama, near Carthage in North Africa. Hannibal's defeat decided the outcome of the Second **Punic War** and established Rome as the undisputed champion of the Mediterranean.

Zeebrugge raid
A daring British attack, in April 1918, on the German naval base at Zeebrugge, a Belgian coastal town in the province of West Flanders, during **World War I**.

Zeebrugge was occupied by the Germans in 1914 and developed as a major **U-boat** and torpedo-boat base. It was frequently bombarded by British warships and bombed from the air but the Germans built large concrete shelters, which were impervious to the bombs of the time. On 23 April 1918, a party of Royal Marines who landed from HMS *Vindictive* raided the base and put it out of action for the rest of the war.

Zeppelin, Count Ferdinand von (1838–1917)
German designer of the rigid-framed **airship** used by Germany in **World War I**. The first Zeppelin that the Count flew, the LZ-1, was 128m/420 ft long with a top speed of 27kph/17mph. The much improved LZ-3, launched in 1906, was adopted by the military.

The Zeppelin L-59 could fly 6,800 km/ 4,200 mi non-stop at over 95 kph/50 mph.

In **World War I**, Germany used Zeppelins to patrol the North Sea, to scout enemy ships and to carry out strategic bombing. However, many Zeppelins were shot down or were lost through accidents or bad weather.

Zhukov, Georgi Konstantinovich (1896–1974)

Soviet field marshal and **World War II** hero. A furrier's apprentice of peasant stock, he was conscripted into the tsar's army before joining the Bolsheviks and fighting in the 1917 Revolution and Russian Civil War (1918–21). He attended the Frunze military academy and in 1936 was sent by Stalin to monitor the **civil war** in Spain. In 1939 he commanded the **Red Army** against the Japanese in Mongolia. During **World War II**, he defeated the Germans at **Stalingrad** and raised the Siege of Leningrad. He attacked and captured Berlin at the end of the war, and became commander of the Soviet zone in occupied Germany, and later Defence Minister.

Appendix

Noteworthy Military and Naval Commanders

Abd el-Krim Mahommed ibn (1882–1963) Moroccan guerrilla leader who fought against the French and the Spanish

Akbar (1542–1605) Mogul emperor of India

Alexander Harold Rupert Leofric, 1st Viscount Alexander (1891–1969) British field marshal who commanded British forces in Middle East and Italy during World War II

Alexander the Great (356–323 BC) king of Macedonia; military genius who defeated the Persian Empire

Allen Ethan (1739–1789) US soldier and politician in the American Revolution

Allenby Edmund Henry Hynman, 1st Viscount Allenby (1861–1936) British field marshal in World War I who fought a victorious Middle East campaign

Alvarez de Toledo Fernando, Duke of Alva (1507–1582) Spanish general who fought campaigns in Hungary, Germany, North Africa, Italy, and the Netherlands

Anders Wladyslaw (1892–1970) Polish general who commanded the 2nd Polish Corps in the Italian campaign during World War II

Arnold Benedict (1741–1801) US turncoat soldier in the American Revolution

Arnold Henry ('Hap') (1861–1950) US aviator and general in World War II; head of the Army Air Corps

Attila the Hun (c. 406–453) king of the Huns; defeated the Romans at Chalons in AD 451, invaded Italy in AD 452

Auchinleck Sir Claude (1884–1891) British field marshal in World War II who commanded British forces in the Middle East

Babur (1483–1530) founder Mogul emperor who conquered northern India

Badoglio Pietro (1871–1956) Italian general during World War II who deposed Mussolini in July 1943

Belisarius (c. 505–565) Byzantine general; fought campaigns against the Vandals, Ostrogoths, Persians, and Kotrigur Huns

Billiere Peter de la (1935–) British general who commanded British forces during the Gulf War (1990–91)

Blücher Gebhard Leberecht von (1742–1819) Prussian general who fought against Napoleon at Waterloo

Bolivar Simon (1783–1830) Venezuelan soldier and political leader who eventually liberated Colombia in 1819

Bouillon Godfrey de (c. 1060–1100) leader of the first Crusade

Bradley Omar Nelson (1893–1981) US army general in World War II who fought in North Africa and commanded the 12th Army Group in Europe

Brian Boru (941–1014) high king of Ireland

Brusilov Aleksei Alekseevich (1853–1926) Russian general in World War I

Bülow Karl von (1846–1921) German general in World War I

Burgoyne John (1722–1792) British general in the American Revolution who was defeated at Saratoga in 1777

Cadorna Count Luigi (1850–1928) Italian general in World War I

Caesar Caius Julius (c. 100–44 BC) Roman emperor who conquered Gaul and defeated Pompey at Pharsalus in 48 BC

Charlemagne (Charles the Great) (742–814) king of the Franks and founder of the Holy Roman Empire

Charles XII (1682–1718) king of Sweden who led his country to defeat in the Great Northern War 1700–21

Chi'in Shih-huang ti (259–210 BC) Chinese emperor who united the seven rival kingdoms of China by military conquest (230–221 BC)

Churchill John, 1st Duke of Marlborough (1650–1722) English general in the War of Spanish Succession; victor at the battle of Blenheim in 1704

Chu Teh (1886–1976) commander of Chinese communist forces during the anti-Japanese War (1937–45), and Chinese Civil War (1945–49)

Clark Mark Wayne (1896–1984) US general who served in Italy during World War II and fought in the Korean War

Clausewitz Karl Maria von (1780–1831) Prussian general and military theorist

Clive Robert, Baron (1725–1774) British general who was victor at Plassey in 1757

Cordoba Gonzales de (1453–1515) Spanish general who defeated the French to secure Naples for Spain in 1503

Cornwallis Charles, 1st Marquess (1738–1805) British general in the American Revolution who was defeated at Yorktown in 1781

Cortés Hernan (1485–1547) Spanish general who conquered the Aztecs in 1521

Cromwell Oliver (1599–1658) English soldier and Lord Protector; he set up the New Model Army in 1645

Cumberland William Augustus, Duke of (1721–1765) British soldier and victor at Culloden in 1746

Cunningham Andrew Browne, 1st Viscount (1883–1963) British admiral in World War II

Custer George Armstrong (1839–1876) US army general who was killed at Little Big Horn

Cyrus the Great (559–530 BC) great king of Persia who captured Babylon in 539 BC

Darius (I) the Great (548–486 BC) king of Persia

Dayan Moshe (1915–1981) Israeli general and politician who was largely responsible for his country's victory in the 1967 Six Day War

Dearborn Henry (1751–1829) US army general in the American Revolution

Dönitz Karl (1891–1980) German admiral in World War II who developed the U-Boat 'Wolf Pack' strategy

Doolittle James ('Jimmy') Harold (1896–1993) US Air Force Commander in World War II

Drake Sir Francis (1540–1596) English sailor and victor against the Spanish Armada in 1588

Dudley Robert, Earl of Leicester (c. 1532–1588) English soldier and favourite of Queen Elizabeth I

Edward III (1312–1377) king of England and victor at Crécy, France, in 1346

Edward Plantagenet 'the Black Prince' (1330–1376) Prince of Wales and victor at Tours in 1356

Eisenhower Dwight David ('Ike') (1890–1969) US general who became Supreme Allied Commander in World War II and president of the USA (1952–60)

El Cid Rodrigo Diaz de Vivar (c. 1043–1099) Spanish general who defeated the Moors at Valencia in 1094

Eugen Prince of Savoy-Carignon (1663–1736) French-born soldier and diplomat in the War of Spanish Succession

Fabius Maximus Verrucosus Quintus ('Cunctator', The Delayer) (d. 203 BC) Roman general who fought against Hannibal

Falkenhayn Erich von (1861–1922) German general during World War I who ordered the bloody assault on Verdun in 1916

Farnese Alessandro, Duke of Parma (1545–1592) Spanish general in the Netherlands

Farragut David Glasgow (1801–1870) US (Union) admiral in the American Civil War who won victory at Mobile Bay in 1864

Foch Ferdinand (1851–1929) French general in World War I, he became Supreme Allied Commander in 1918

Frederick I ('Barbarossa') (c. 1123–1190) emperor of the Holy Roman Empire

Frederick (II) the Great (1713–1786) Prussian king who fought the Seven Years' War

French John Denton Pinkstone, 1st Earl

(1852–1925) British field marshal in World War I who commanded the British Expeditionary Force (1914–15)

Gallieni Joseph-Simon (1849–1916) Marshal of France who dispatched troops in Paris taxis to the Battle of the Marne

Gamelin Maurice (1872–1958) French military commander, and chief of staff (1935–40)

Garibaldi Giuseppe (1807–1882) Italian general and patriot who fought to unify Italy

Gaulle Charles André Joseph Marie de (1890–1970) commander of Free French Forces during World War II, who later became President of France

Genghis Khan (Temujin) (1162–1227) fearsome soldier and founder of the Mongol Empire

Geronimo (*c.* 1829–1909) American Indian (Apache) warband leader

Ghormley Robert Lee (1883–1953) US admiral in World War II

Giap Vo Nguyen (1910–) Vietnamese communist military leader who fought against the French and the Americans

Glyn Dwr Owain (*c.* 1354–*c.* 1416) Welsh patriot ('Prince of Wales') and soldier

Gordon Charles George ('Gordon of Khartoum') (1833–1885) British general killed at Khartoum in the Sudan in 1885

Gort John (1886–1946) British field marshal in World War II who commanded the British Expeditionary Force (1939–40)

Grant Ulysses Simpson (born Hiram Ulysses Grant) (1822–1885) US (Union) general in the American Civil War who later became president of the USA

Greene Nathanael (1742–1786) US general in the American Revolution

Guderian Heinz (1888–1956) German general during World War II who developed use of armoured forces

Guesclin Bertrand de (*c.* 1320–1380) French general in the Hundred Years' War

Guevara Ernesto ('Che') (1928–1967) Argentine-born revolutionary who fought in Cuba (1956–59) and was killed in Bolivia

Gustavus II Adolphus (1594–1632) king of Sweden who fought in the Thirty Years' War

Haig Douglas, 1st Earl (1861–1928) British field marshal, commander in chief of the British Expeditionary Force in World War I

Halsey William Frederick (1882–1959) US admiral in World War II who fought in the South Pacific

Hannibal (247–182 BC) Carthaginian soldier and victor at Cannae in 216 BC

Harris Arthur ('Bomber Harris') (1892–1984) British Air Commander during World War II who advocated strategic bombing of Germany

Henry V (1387–1422) king of England and victor at Agincourt, France, in 1415

Hildeyoshi Toyotomi (1536–1598) Japanese general

Hindenburg Paul Ludwig von Beneckendorf von (1847–1934) German general in World War I, he was elected President of the Weimar Republic in 1925

Hitler Adolf (1889–1945) German military and political leader during World War II

Hodges Courtney Hicks (1887–1966) US general who commanded the First Army in Europe during World War II

Homma Masaharu (1888–1946) Japanese general during World War II who invaded the Philippines

Horrocks Sir Brian (1895–1984) British general and corps commander in Africa and Europe during World War II

Hoth Hermann (1885–1971) German general who fought in Russia (1941–43) during World War II

Houston Sam(uel) (1793–1863) US general in the Mexican War of 1836 that gained independence for Texas

Howe Richard, 1st Earl (1726–1799) British admiral in the American Revolution

Jackson Thomas Jonathan ('Stonewall') (1824–1863) US Confederate general in the American Civil War

Jeanne d'Arc ('La Pucelle') (1412–1431) French peasant-girl patriot in the Hundred Years' War

Jellicoe John Rushworth, 1st Earl (1859–1935) British admiral in World War I who commanded British naval forces at the Battle of Jutland in 1916

Jervis John, 1st Earl of St Vincent (1735–1823) British admiral during the Napoleonic Wars

Jiang Jie Shi (1887–1975) nationalist Chinese general and leader who took part in the revolution of 1911

Joffre Joseph Jacques Cesaire (1852–1931) Marshal of France, and commander in chief of French Forces (1914–16)

Jones John Paul (1747–1792) US sailor in the American Revolution

Kemal Mustafa (known as Atatürk, 'Father of Turks') (1881–1938) Turkish general and statesman who fought in World War I and against the Greeks

Kesselring Albert (1885–1960) German field marshal, commander of German forces in Africa and Italy during World War II

King Ernest Joseph (1878–1956) US admiral and commander in chief and chief of naval operations during World War II

Kinkaid Thomas Cassin (1888–1972) US admiral who fought at Coral Sea, Midway, Guadalcanal, Leyte, and Luzon during World War II

Kitchener Horatio Herbert, 1st Earl (1850–1916) British field marshal who fought in the Sudan and the Boer War; he raised a volunteer army in World War I

Kleist Paul Ewald von (1881–1954) German field marshal best known for campaigns in France and Russia during World War II

Koga Mineichi (1885–1944) Japanese admiral in World War II who became commander in chief on the death of Yamamoto

Konev Ivan Stepanovich (1897–1973) marshal of the USSR during World War II

Kosciuszko Tadeusz Andrzej Bonawentura (1746–1817) Polish general who fought for Poland's independence and later in the American Revolution

Kutuzov Mikhail Larionovich Golenishchev (1745–1813) Russian general in the Napoleonic Wars who was defeated at Austerlitz but successfully countered Napoleon's invasion of Russia in 1812

Lafayette Joseph Paul Roch Yves Gilbert Motier (1757–1834) French general in the American Revolution

Lattre de Tassigny Jean Marie Gabriel de (1889–1952) French general in World War II and Indochina

Lawrence T(homas) E(dward) ('Lawrence of Arabia') (1888–1935) British military leader, soldier, and writer, best known for victories at Aqaba (1917) and Damascus (1918) during World War I

Lee Robert Edward (1807–1870) US Confederate general in the American Civil War

Leigh-Mallory Trafford (1892–1944) British air commander during World War II, he commanded Allied air forces during the Normandy invasion

LeMay Curtis Emerson (1906–1990) US Air Force general in World War II

Ludendorff Erich (1865–1937) German general in World War I, victor at Tannenberg (1914); his 1918 offensive ended in failure

MacArthur Douglas (1880–1964) US general in the Pacific during World War II, and in the Korean War

Mannerheim Carl Gustaf Emil, Baron von (1867–1951) Finnish general in World War II

Manstein Fritz Erich von (1887–1973) German field marshal during World War II, brilliant strategist who devised the plan to defeat France

Manteuffel Hasso-Eccard von (1897–1978) German general who fought campaigns in Africa, Russia, and the Ardennes during World War II

Mao Tse-tung (1893–1976) Chinese communist war leader who fought against the Japanese (1937–45)

Marshall George Catlett (1880–1959) US general, chief of staff of US Armed Forces throughout World War II

Martel Charles (c. 689–741) Carolingian general who fought against the Muslims

Maurice of Nassau Prince of Orange (1567–1625) Dutch general who successfully defended the United Netherlands against Spain following the Dutch Revolt

Milch Erhard (1892–1972) German field marshal who was largely responsible for the setting up of the Luftwaffe

Mitscher Marc Andrew (1887–1947) US admiral in World War II who commanded carrier forces in the Pacific

Model Otto Moritz Walter (1891–1945) German field marshal during World War II who became known as the 'Führer's Fireman'

Moltke Helmuth Karl Bernhard von ('the Elder', 1800–1891) German general who designed the General Staff System and defeated France in the Franco-Prussian War

Moltke Helmut von ('the Younger', 1848–1916) German general during World War I whose flawed execution of Schlieffen's Plan caused Germany's defeat

Montcalm Louis Joseph (1712–1759) French general defeated at Québec in 1759

Montgomery Bernard Law, 1st Viscount of Al 'Alamayn (1887–1976) British general in World War II and victor at Al 'Alamayn, Egypt, in 1942

Moore (John) Jeremy (1928–) British general who commanded British land forces during the Falklands War in 1982

Mountbatten Louis Francis Albert Nicholas, Earl (1900–1979) British admiral and pioneer of combined operations during World War II who also fought the Japanese in Burma

Nagumo Chuichi (1886–1944) Japanese admiral during World War II who commanded carrier forces during Pearl Harbor and Midway

Napoleon I (Bonaparte) (1769–1821) emperor of France and military genius

Nelson Horatio, 1st Viscount (1758–1805) British admiral in the Napoleonic Wars and victor at Trafalgar in 1805

Ney Michel, Prince of Moscow (1769–1815) French general in the Napoleonic Wars

Nimitz Chester William (1885–1966) US admiral and commander in chief of the Pacific Fleet and Pacific Areas during World War II

Nivelle Robert-Georges (1856–1924) French general during World War I whose 1917 offensive was a costly failure

O'Connor Richard (1889–1981) British general during World War II who defeated Italian forces in the Western desert in 1940

Patton George Smith Jr (1885–1945) US general during World War II, famous for his breakout from the Normandy bridgehead

Perry Matthew Calbraith (1794–1858) US admiral who opened up Japan to trade

Pershing John Joseph ('Black Jack') (1860–1948) US general who commanded the American Expeditionary Force in France during World War I

Pétain Henri Philippe Omer (1856–1951) French general in World War I and leader of Vichy France in World War II

Peter (I) the Great (1672–1725) Russian tsar who successfully fought the Great Northern War against Sweden (1700–21)

Phillip II (c. 382–336 BC) king of the Macedonians who defeated the Greeks at Chaeronea in 338 BC

Pontiac (c. 1720–1769) American Indian war-chief (Ottawa tribe) who allied with the French during the French–Indian War

Powell Colin (Luther) (1937–) US general who, as chair of the joint chiefs of staff 1989–93, was responsible for the overall administration of the Allied Forces in Saudi Arabia during the Gulf War.

Richard (I) the Lionheart (1157–1199) king of England and crusader

Rickover Hyman G (1900–1986) US admiral who pioneered the use of nuclear technology to power warships

Ridgway Matthew Bunker (1895–1993) US general in World War II and the Korean War

Robert (I) the Bruce (1274–1329) king of Scotland and victor at Bannockburn in 1314

Roberts Frederick Sleigh, 1st Earl (1832–1914) British general in India

Rokossovski Konstantin Konstantinovich (1896)–1968) Soviet general in World War II

Rommel Erwin Johannes Eugen (1891–1944) German field marshal during World War II, best known for his North African campaign

Rundstedt Karl Rudolf Gerd von (1875–1953) German field marshal during World War II who fought successful campaigns in Poland and France

Ruyter Michiel Adriaanzoon de (1607–1676) Dutch admiral

Saladin (or Salah-al-din) (1138–1193) Muslim military leader who defeated the Crusaders at Hatlin and took Jerusalem in 1187

San Martin José de (1778–1850) Argentine-born Latin American general

Saxe Herman Maurice (1696–1750) German-born French general in the War of Austrian Succession

Seeckt Hans von (1866–1936) German general who reformed the army between the wars and developed the Blitzkrieg tactics

Sherman William Tecumseh (1820–1891) Union general during the American Civil War

Schwarzkopf H(erbert) Norman ('Stormin' Norman') (1934–) US general who commanded the multinational coalition forces to victory in the Gulf War

Scipio Africanus Publius Cornelius (237–183 BC) Roman general who defeated Hannibal at Zama in 202 BC and conquered Carthage

Scott Winfield (1786-1866) US general who modernized the army along European lines and fought in the Mexican War

Shaka (1787–1828) founder of the Zulu Empire in southern Africa

Sharon Ariel ('Arik') (1928–) Israeli general and politician

Sheridan Philip Henry (1831–1888) US Union general in the American Civil War

Sikorski Wladyslaw (1881–1943) Polish general and commander of Free Polish forces during World War II

Sims William Sowden (1858–1936) US admiral in World War I

Skorzeny Otto (1908–1975) German commander whose unit successfully rescued Mussolini and fought during the Battle of the Bulge

Slim William Joseph, 1st Viscount (1891–1970) British general who commanded the 14th Army in the successful Burma campaign during World War II

Smith Holland McTyeire ('Howlin Mad') (1882–1967) tough US marine general who fought throughout the Pacific during World War II

Smuts Jan Christiaan (1870–1950) South African military commander who fought with the British during both World Wars

Somerset Fitzroy James Henry, 1st Baron Raglan (1788–1855) British general in the Crimean War

Spaatz Carl ('Tooey') (1891–1974) US Air Force general in World War II

Spruance Raymond Amos (1886–1969) US admiral who fought in the Pacific during World War II

Stilwell Joseph Warren (1883–1946) US general who fought in South East Asia during World War II

Stirling David (1915–1990) British soldier who created the Special Air Service (SAS) during World War II

Stuart James Ewell Brown ('Jeb') (1833–1864) US Confederate general in the American Civil War

Student Kurt (1890–1978) German general who pioneered the use of airborne forces during World War II

Suleiman (I) the Magnificent (1494–1566) Ottoman sultan who was victorious at Mohács in 1526 and besieged Vienna in 1529

Sun Tzu (dates unknown, $c.$ 400 BC) ancient Chinese military theorist who wrote *The Art of War*

Suvorov Aleksandr Vasil'evich (1730–1800) Russian general in the Seven Years' War and French Revolutionary War who defeated the French at Adda (1799)

Terauchi Hisaichi (1879–1945) Japanese general who directed the invasions of Indochina, Siam, Malaya, and Java during World War II

Tilly Johann Tserclaes, Graf von (1559–1632) Flemish general in the Thirty Years' War

Timur-I-Leng (Tamerlane) (1336–1405) Mongol warrior who conquered Central Asia and founded the Timurid dynasty

Tito (Josip Broz) (1892–1980) Yugoslav guerrilla leader in World War II

Togo Count Heihachiro (1849–1934) Japanese admiral and victor at Tsushima in 1905

Tojo Hideki (1884–1948) Japanese general and politician in World War II

Turenne Henri de la Tur d'Auvergne, Vicomte de (1611–1675) French general in the Thirty Years' War

Turner Richmond Kelly (1885–1961) US admiral who served in the Pacific in World War II

Vandegrift Alexander Archer (1887–1973) US marine general in World War II

Vasilevsky Aleksander Mikhailovich (1895–1977) Soviet general during World War II, best known for victories at Stalingrad and Kursk

Vatutin Nikolai (1901–1944) Soviet general who fought successfully at Stalingrad and Kursk, and retook Kiev during World War II

Vauban Sébastien le Prestre de (1633–1707) French military engineer and designer of fortifications

Villa Francisco ('Pancho') (1877–1923) Mexican revolutionary general

Villeneuve Pierre Charles Jean Baptiste de (1763–1806) French admiral defeated at Trafalgar in 1805

Wallenstein Albrecht Eusebius Wenzel von (1583–1634) Czech-born general in the Thirty Years' War

Washington George (1732–1799) US general during the American Revolution and first President of the USA

Wellesley Arthur, 1st Duke of Wellington (1769–1852) British general in the Napoleonic Wars and victor at Waterloo in 1815

Westmoreland William Childs (1914–) US general in the Vietnam War

Wingate Orde Charles (1903–1944) British general who pioneered the use of deep penetration 'Chindit' units in Burma

Wolfe James (1727–1759) British general and victor at Québec in 1759

Woodward John (Sandy) (1931–) British admiral who commanded the British Task Force during the Falklands War in 1982

Xenophon (c. 435–c. 354 BC) Greek soldier and writer

Xerxes I (c. 520–465 BC) king of Persia who was defeated at Thermopylae by Leonidas in 480 BC and in the naval battle at Salamis in 480 BC

Yamamoto Isoroku (1883–1943) Japanese admiral who was the architect of the Pearl Harbor battle in 1941

Yamashita Tomoyuki (1888–1946) Japanese general during World War II who defeated British forces in Malaya and Singapore

Zhukov Georgi Konstantinovich (1896–1974) Marshal of the USSR during World War II, he defended Moscow and captured Berli

Navy: Chronology

5th century BC   Naval power is an important factor in the struggle for supremacy in the Mediterranean; for example, the defeat of Persia by Greece at Salamis.

311 BC   The first permanent naval organization is established by the Roman Empire with the appointment of navy commissioners to safeguard trade routes from pirates and eliminate the threat of rival sea power.

878 AD   Alfred the Great of England overcomes the Danes with a few king's ships, plus ships from the shires and some privileged coastal towns.

12th century   Turkish invasions end Byzantine dominance.

13th century   The first French royal fleet is established by Louis IX. His admirals come from Genoa.

1339–1453   During the Hundred Years' War there is a great deal of cross-Channel raiding by England and France.

16th century   Spain builds a large navy for exploration and conquest in the early part of the century. In England, building on the beginnings made by his father Henry VII, Henry VIII raises a force that includes a number of battleships, such as the 'Mary Rose', creates the long-enduring administrative machinery of the Admiralty, and, by mounting heavy guns low on a ship's side, revolutionizes strategy by the use of the "broadside". Elizabeth I encourages Drake, Frobisher, Hawkins, Raleigh, and other navigators to enlarge the empire.

1571	The Battle of Lepanto is one of the last to be fought with galleys, or oar-propelled ships.
1588	The defeat of the Spanish Armada begins the decline of the sea power of Spain.
17th century	There is a substantial development in naval power among the powers of northern Europe; for example, in the Netherlands, which then founds an empire in the Americas and the East; France, where a strong fleet is built up by Richelieu and Louis XIV that maintains the links with possessions in India and North America; and England, comparatively briefly under Cromwell. In the late 17th century the British overtake the Dutch as the leading naval power.
1775–83	The US navy grows out of the coastal colonies' need to protect their harbours during the American Revolution, as well as the need to capture British war supplies. In late 1775 Washington prepares five schooners and a sloop, crewed with army personnel, and sends them to prey on inbound supply vessels. By the time of the Declaration of Independence 1776 these are augmented by armed brigs and sloops from the various colonies. The hero of the period is John Paul Jones.
1805	Effectively reorganized by Pitt in time for the French Revolutionary Wars, the British Royal Navy under Nelson wins a victory over the French at Trafalgar, which ensures British naval supremacy for the rest of the 19th century.
19th century	The US fleet is successful in actions against pirates off Tripoli 1803–05 and the British navy 1812–14, and rapidly expands during the Civil War and again for the Spanish-American War in 1898.
World War I	Britain maintains naval supremacy in the face of German U-boat and surface threats.
1918–41	Between the wars the US fleet is developed to protect US trade routes, with an eye to the renewed German threat in the Atlantic and the danger from Japan in the Pacific.
1950s	After World War II the US fleet emerges as the world's most powerful.
1962	The Cuban missile crisis (when the USA forces the removal of Soviet missiles from Cuba) demonstrates the USSR's weakness at sea and leads to its development under Admiral Sergei Gorshkov.
1980s	The Soviet fleets (based in the Arctic, Baltic, Mediterranean, and Pacific) continue their expansion, rivalling the combined NATO fleets. The new pattern of the Soviet navy reflects that of other fleets: over 400 submarines, many with ballistic-missile capability, and over 200 surface combat vessels (mostly of recent date) including helicopter carriers, cruisers, destroyers, and escort vessels. The USA maintains aircraft-carrier battle groups and recommissions World War II battleships to give its fleet superior firepower, as well as smaller support vessels.
1990s	With the end of the Cold War, the powerful Soviet navy begins to fall apart. The Black Sea fleet is split between Russia and the Ukraine, the Baltic fleet loses its main ports to the newly independent Baltic states, and the Northern and Far Eastern fleets curtail their operations for lack of resources. All leading Western nations, including the USA, announce plans to reduce in size.

Arms Control Agreements

Treaty	Dates	Parties	Details
Agreements (Nuclear)			
Antarctic Treaty	signed 1 December 1959; entered into force 23 June 1961	37 nations	prohibited any use of Antarctica for military purposes; specifically prohibited nuclear testing and nuclear waste
Partial Test Ban Treaty	signed 5 August 1963; entered into force 10 October 1963	125 nations; an additional 10 nations have signed the treaty, but have not ratified it	prohibited nuclear weapon tests in outer space, in the atmosphere, and underwater
Outer Space Treaty	signed 27 January 1967; entered into force 10 October 1967	96 nations; an additional 27 nations have signed the treaty, but have not ratified it	prohibited nuclear weapons in Earth orbit and outer space
Treaty of Tlatelolco	signed 14 February 1967; entered into force 22 April 1968	32 nations; the USA, the UK, France, Russia, China, and the Netherlands have signed the relevant protocols to the treaty	prohibited nuclear weapons in Latin America and required safeguards on facilities
Non-proliferation Treaty	signed 1 July 1968; entered into force 5 March 1970; renewed indefinitely in 1995	188 nations; all nations that have signed have also formally adopted the treaty; the principal non-signatories are Israel, India, and Pakistan	divided the world into nuclear and non-nuclear weapon states based on status in 1968; obliged non-nuclear states to refrain from acquiring nuclear weapons and to accept International Atomic Energy Agency (IAEA) safeguards on their nuclear energy facilities; obliged nuclear states to refrain from providing nuclear weapons to non-nuclear states, to assist in the development of nuclear energy in non-nuclear states, and to work toward global nuclear disarmament
Seabed Arms Control Treaty	signed 11 February 1971; entered into force 18 May 1972	94 nations; an additional 20 nations have signed the treaty, but have not ratified it	prohibited placement of nuclear weapons on the seabed and ocean floor beyond a 19 km/12 mi coastal limit

Agreement	Signed/Entered into Force	Parties	Description
Accident Measures Agreement	signed and entered into force 30 September 1971	USA and USSR (Russia)	measures to prevent accidental nuclear war, including agreement to notify each other of planned missile launches and the detection of unidentified objects
SALT I—Interim Agreement	signed 26 May 1972; entered into force 3 October 1972	USA and USSR (Russia); USA dropped commitment to agreement in 1986	froze at existing levels the number of strategic ballistic missile launchers on each side; permitted an increase in submarine-launched ballistic missiles (SLBMs) up to an agreed level only with the dismantling or destruction of a corresponding number of older intercontinental ballistic missile (ICBM) or SLBM launchers
ABM Treaty	signed 26 May 1972; entered into force 3 October 1972	USA and USSR (Russia)	prohibited nationwide anti-ballistic missile (ABM) defences, limiting each side to 2 deployment areas (1 to defend the national capital and 1 to defend an ICBM field) of no more than 100 ABM launchers and interceptor missiles each
Prevention of Nuclear War Agreement	signed and entered into force 22 June 1973	USA and USSR (Russia)	agreed to make the removal of the danger of nuclear weapons and their use an "objective of their policies"; committed to consult with each other in case of danger of nuclear confrontation between them or other countries
ABM Protocol	signed 3 July 1974; entered into force 24 May 1976	USA and USSR (Russia)	limited each side to a single ABM deployment area
Threshold Test Ban Treaty	signed 3 July 1974; entered into force 11 December 1990	USA and USSR (Russia)	limited nuclear tests on each side to a 150 kiloton threshold
Peaceful Nuclear Explosions Treaty	signed 28 May 1976; entered into force 11 December 1990	USA and USSR (Russia)	an agreement not to carry out any underground nuclear explosion for peaceful purposes having a yield exceeding 150 kilotons, or any group explosion having an aggregate yield exceeding 150 kilotons
US–International Atomic Energy Agency Safeguards Agreement	signed 18 November 1977; entered into force 9 December 1980	USA and the IAEA	agreed to apply IAEA safeguards in designated facilities in the USA

Treaty	Signed	Parties	Provisions
SALT II Treaty	signed 18 June 1979	USA and USSR (Russia); USA dropped commitment to the agreement in 1986	never ratified, but both nations pledged to comply with the treaty's provisions; provided for broad limits on strategic offensive nuclear weapons systems, including equal numbers of strategic nuclear delivery vehicles, and restraints on qualitative developments that could threaten future stability; superceded by START in 1991
Convention on the Physical Protection of Nuclear Material	signed 3 March 1980; entered into force 8 February 1987	58 nations; an additional 7 nations have signed the treaty, but have not ratified it	provided for certain levels of physical protection during international transport of nuclear materials
Treaty of Rarotonga	signed 6 April 1985; entered into force 11 December 1986	12 nations are party to the treaty; all 5 declared nuclear weapons states have signed the relevant protocols to the treaty; the USA signed the protocols on 25 March 1996, but has yet to ratify the treaty	established a nuclear-weapons-free zone in the South Pacific
Missile Technology Control Regime	informal association formed in April 1987	28 nations	established export control guidelines and annexed listing nuclear-capable ballistic missile equipment and technologies that would require export licenses
Nuclear Risk Reduction Centres	signed and entered into force 15 September 1987	USA and USSR (Russia)	established centres designed to reduce the risk of nuclear war
Intermediate-Range Nuclear Forces Treaty	signed 8 December 1987; entered into force 1 June 1988	USA and USSR (Russia)	eliminated all US and Soviet ground-launched ballistic and cruise missiles with ranges of 500–5,500 km/310–3,418 mi, their launchers, associated support structures, and support equipment
Ballistic Missile Launch Notification Agreement	signed and entered into force 31 May 1988	USA and USSR (Russia)	agreement to notify each other, through nuclear risk reduction centres, no less than 24 hours in advance, of the planned date, launch area, and area of impact for any launch of an ICBM or SLBM

Treaty	Signed/Force	Parties	Details
START I Treaty	signed 31 July 1991; entered into force 5 December 1994	USA and USSR (Russia)	Russia formally accepted all of the USSR's arms control treaty obligations after the dissolution of the USSR. Because strategic nuclear weapons affected by the START I Treaty remained on the soil of Ukraine, Kazakhstan, and Belarus after the USSR's demise, an additional agreement was necessary to guarantee their removal; this was assured under the Lisbon protocol of May 1992. START established limits on deployed strategic nuclear weapons, and required the USA and Russia to make phased reductions in their offensive strategic nuclear forces over a 7 year period
START II Treaty	signed 3 January 1993; ratified by USA 26 January 1996	USA and Russia; still awaiting ratification by Russia. Both sides committed to negotiating START III upon Russia's ratification of START II	limited each side to 3,000–3,500 deployed strategic nuclear weapons each by the year 2003; banned multiple-warhead land-based missiles
Treaty of Bangkok	signed 15 December 1995; entered into force 27 March 1977	10 nations; none of the acknowledged nuclear powers have signed the relevant protocols. Brunei, Cambodia, Indonesia, Laos, Malaysia, Myanmar, Singapore, Thailand, and Vietnam have now ratified the treaty; the Philippines have signed but not ratified it	established a nuclear-weapons-free zone in Southeast Asia
Treaty of Pelindaba	signed 11 April 1996; the USA signed on 11 May 1996. Treaty open to signature by all states of Africa. It will enter into force upon the 28th ratification	54 signatories; the 5 acknowledged nuclear weapons states have signed the relevant protocols to the treaty and 8 states have now ratified it	established a nuclear-weapons-free zone in Africa

Treaty	Signed	Parties	Description
Comprehensive Test Ban Treaty	signed 24 September 1996	as of January 1999, 177 nations had signed the treaty, but only 26 nations had formally ratified it—to enter into force, the 44 nations that have nuclear electrical-generating and research reactors are required to sign and ratify the treaty; included in the 44 are India, Pakistan, and North Korea, none of which have signed the treaty. On 6 April 1998, Britain and France became the first of the five main nuclear powers to ratify the treaty	banned all explosive nuclear tests
Joint Statement on Parameters on Future Reductions in Nuclear Forces	signed 21 March 1997	USA and Russia	USA and Russia have agreed that negotiations for START III will begin as soon as START II enters into force

Agreements (Non-Nuclear)

Treaty	Signed	Parties	Description
Geneva Protocol	signed 17 June 1925; entered into force 8 February 1928; ratified by USA 22 January 1975	132 nations	prohibited in war the use of asphyxiating or poisonous gas and liquids and all bacteriological (biological) methods of warfare; essentially a "no first use" agreement, because many signatories ratified it with the reservation that it would cease to be binding if an enemy failed to observe the prohibitions

Biological and Toxic Weapons Convention	signed 10 April 1972; entered into force 26 March 1975	142 nations; 18 nations signed but not ratified	prohibited development, production, stockpiling, acquisition, or retention of biological agents not associated with peaceful uses, biological or toxin weapons, and their delivery systems; required destruction or diversion of all prohibited agents to peaceful purposes within 9 months after entry into force; prohibited transfer of, or assistance to manufacture, prohibited agents
Incidents at Sea Agreement	signed and entered into force 25 May 1972	USA and USSR (Russia)	agreement on the prevention of incidents on and over the high seas, including steps to avoid collision and the avoidance of manoeuvres in areas of heavy sea traffic
Environmental Modification Prevention	signed 18 May 1977; entered into force 5 October 1978	64 nations; an additional 17 nations have signed the treaty, but have not ratified it	prohibited deliberate manipulation of natural processes for hostile or other military purposes
Protocol (I) Additional to the 1949 Geneva Convention and Relating to the Protection of Victims of International Conflict	signed 12 December 1977; entered into force 7 December 1978	148 nations; in addition Cambodia acceded to Protocol on 14 January 1998, and UK ratified Protocol on 28 January 1998	Protocol prohibits the use of weapons or means of warfare which cause superfluous injury or unnecessary suffering
Convention on Certain Conventional Weapons (The Inhumane Weapons Convention)	opened for signature 10 April 1981; entered into force 2 December 1983; signed by the USA 2 April 1982 and ratified 24 March 1995	72 nations; 10 nations have signed the convention but not ratified it; 29 nations have signed the amended Protocol II; 34 nations have signed Protocol IV	banned the use of non-metallic fragmentation weapons and blinding lasers, and imposed restrictions on the use of land mines and incendiary weapons
Australia Group	informal association formed in 1984 in response to chemical weapons use in the Iran–Iraq War	30 nations	established voluntary export controls on certain chemicals

Treaty	Signing/Entry	Parties	Description
Stockholm Conference	signed and entered into force 19 September 1986	54 nations	established security and confidence-building measures designed to increase openness and predictability with regard to military activities in Europe
US–Soviet Bilateral Memorandum of Understanding	signed and entered into force September 1989	USA and USSR (Russia)	agreement to exchange data on size, composition, and location of chemical weapon stockpiles and storage, production, and destruction facilities, and to conduct reciprocal visits and both routine and challenge inspections
US–Soviet Bilateral Destruction Agreement	signed June 1990; has not yet entered into force	USA and USSR (Russia); currently abandoned by Russia	agreement to stop producing chemical weapons and to reduce their stockpiles to no more than 5,000 agent tons, with destruction to begin by 31 December 1992 and to be completed by 31 December 2002
Conventional Forces in Europe (CFE) Treaty	signed 19 November 1990; entered into force 9 November 1992	30 nations	established limits on the numbers of tanks, armoured combat vehicles, artillery combat aircraft, and helicopters in Europe
Open Skies Treaty	signed 24 March 1992	27 nations; the USA and 22 other nations have ratified the treaty; ratifications by Russia, Belarus, Ukraine, Kyrgyztan, and Georgia are still required for entry into force	permitted unarmed reconnaissance flights designed to promote military transparency and confidence among former NATO and Warsaw Pact adversaries
CFE 1A Treaty	signed 10 July 1992; entered into force 9 November 1992	30 nations	established limits on the number of troops in Europe
Chemical Weapons Convention	opened for signature 13 January 1993; entered into force 29 April 1997	169 nations; 121 signatories have ratified the treaty	prohibited the development, production, acquisition, stockpiling, retention, transfer, and use of chemical weapons—required declaration of chemical weapons or chemical weapon capabilities, including those primarily associated with commercial use; required destruction of declared chemical weapons and associated production facilities; provided for routine inspections of relevant facilities and challenge inspections upon request

Florence Agreement	signed 14 June 1996, entered into force immediately upon signature	Bosnia-Herzegovina, Bosnian Serb Republic, Croatia, Serbia, and Montenegro	the agreement was negotiated under the Dayton Agreement. It set limits on armaments of the warring parties of the conflict in Bosnia-Herzegovina. It covered heavy conventional weapons including main battle tanks, armoured combat vehicles, heavy artillery (75mm and above), combat aircraft, and attack helicopters
Ottawa Treaty	signed 3–4 December 1997	135 countries; 77 countries had ratified the treaty as at 29 April 1999	banned the production, use, and export of anti-personnel mines; also committed signatories to the destruction of stockpiles of mines within four years, and to clearing existing minefields as well as relieving the suffering of the victims of land mines

Weapons: Chronology

1000 BC — Clubs, hammers, axes daggers, spears, swords, slings, and bows are all common weapons by this time. Armour is also in use for personal protection.

700 BC — The Assyrians develop the ram for battering down walls.

500 BC — The crossbow appears in China.

600 AD — The stirrup, introduced into Europe from the East, gives riders better control and allows the more effective use of lances and swords from the saddle.

668 — First reported use of "Greek Fire", an incendiary compound.

13th century — Gunpowder is brought to the West from China (where it is long in use but only for fireworks).

1232 — First reliable report of the use of rockets by the Chinese.

1242 — First written record of gunpowder.

c.1300 — Guns invented by the Arabs, with bamboo muzzles reinforced with iron.
First mention of firearms, in England and Italy, in the form of small cannon.

1346 — Battle of Crécy, in which gunpowder is probably used in battle for the first time.

1376 — Explosive shells are used in Venice.

1378 — Cannon are mounted in warships by the Venetians.

1400 — Hand firearms—"hand-gonnes"—appear in Europe.

1411 — First mention of the matchlock to ignite hand weapons.

1421 — First record of gunpowder-filled explosive shells fired from cannon.

1450 — Casting of cannon becomes common.

1451 — Mortars, fired at a high angle to drop projectiles on to the enemy, are developed by the Turks.

1500 — The arquebus and hackbut, the first shoulder-fired arms.

1530 — The wheel lock appears, allowing guns to be carried ready for firing, leading to the development of the pistol, a single-handed weapon.

1547 — Earliest flintlocks are developed in Spain.

1570 — Rifled firearms are developed in Germany.

1579 — First use of red-hot shot from cannon.

1596 — Introduction of a wooden, gunpowder-filled fuse to ignite explosive shells so as to burst them in the air over the target.

17th century Widespread use of guns and cannon in the Thirty Years' War and English Civil War.

1625 Introduction of wheeled carriages for field artillery.

1635 The perfected French flintlock mechanism, which becomes the standard ignition method in universal use by 1660.

1647 The French introduce the bayonet.

1750 Introduction of horse artillery, in which the gunners are mounted, giving greater mobility than ordinary horse-drawn artillery where the gunners march alongside their guns.

1776 British adopt the Firguson breech-loading rifle for use during the American Revolution.

1778 The Carronade, a light, short-range gun, is developed for naval use.

1787 Lt Shrapnel begins development of the "spherical case shot" which eventually becomes the "shrapnel" shell.

1792 The Indians under Tipu use rockets against the British at Seringapatam.

1800 Henry Shrapnel invents shrapnel for the British army.

1806 British use the Congreve rocket against Boulogne.

1807 Rev Forsyth patents the percussion principle for ignition of firearms.

1812 Self-contained cartridges for small arms.

1813 British artillery first fires over the heads of assaulting infantry in order to suppress enemy fire until the infantry are close enough to charge.

1814 Development of the percussion cap.

1818 Collier and Wheeler develop a hand-rotated flintlock revolver.

1836 Samuel Colt patents his percussion revolver.

1841 The "Needle Gun", first bolt-action breech-loading rifle with self-contained cartridge, adopted by Prussia.

1846 Invention of guncotton, the first "moder" explosive, by Schonbein.

1849 Beginning of a move to convert muzzle-loading muskets into breech-loaders.

1855 Rifled muzzle-loading cannon adopted in France.

1857 Smith and Wesson patent the first metallic rim-fire cartridge revolver.

1859 Britain adopts Armstrong's rifled breech-loading cannon.

1858 French introduce ironclad warships.

1861 The Gatling gun, first successful mechanical machine gun.

1862 American Civil War introduces breech-loading small arms into military use.

1862 First battle between ironclad warships ('Monitor' and 'Merrimac') and first use in combat of a gun turret on a warship ('Monitor').

1863 TNT discovered by German chemist J Wilbrand.

1864 The British revert to rifled muzzle-loading cannon in order to produce guns of sufficient power to attack ironclad ships.

1865 Invention of smokeless powder by Schultz.

1866 Whitehead's self-propelled naval torpedo.

1866 Alfred Nobel invents dynamite.

1868 General adoption of electrically fired sea mines for harbour protection.

1870 German development of an anti-aircraft gun to shoot at balloons escaping from besieged Paris.

1880 General adoption of breech-loading artillery.

1880 TNT (high explosive) perfected.

1884 Hiram Maxim develops the first automatic machine gun.

1885 The Brennan wire-guided torpedo is adopted in Britain for coastal defence.

1886 France adopts the Lebel rifle, the first bolt-action, magazine repeater of small caliber.

1888 First use of the Maxim machine gun in war.

1890 Bolt action magazine rifles such as the Mauser, Lee-Enfield, Mannlicher, and Krag-Jorgensen models become the universal infantry weapon.

1890 General adoption of cast picric acid as a high explosive filling for artillery shells under various names such as "Lyddite", "Schneiderite".

1893 The Borchardt, first practical automatic pistol.

1897 The French introduce the 75-mm quick-firing gun, the first field artillery piece to have on-carriage recoil control, self-contained

ammunition, and a shield to protect the gunners.
- **1898** Development of the Madsen light machine gun, the first single operator magazine-fed machine gun.
- **1898** First practical submarine.
- **1904** Russo-Japanese war sees the revival of the hand grenade.
- **1906** First successful flight of a Zeppelin airship.
- **1909** Self-propelled anti-aircraft guns are developed in Germany.
- **1912** TNT adopted as a filling for artillery shells in place of picric acid.
- **1911** Aircraft are used for scouting, observation, and bomb-dropping in the Italian-Turkish War.
- **1914** Development of rifle-propelled grenades.
- **1914** First use of firearms from aircraft.
- **1914** Armoured cars in use in Belgium.
- **1915** First use of chemical gas in war: the Germans use tear-gas shells against the Russians at Bolimov.
- **1915** Introduction of synchronized machine guns capable of firing through a spinning propeller without hitting the blades.
- **1915** First use of cloud poison gas in war by the Germans against the British at Ypres.
- **1915** German introduction of the flamethrower.
- **1915** Stokes' trench mortar is developed; it is the precursor of all mortars since then.
- **1916** Introduction of the first sub-machine gun.
- **1916** Tanks are used for the first time by the British at Cambrai.
- **1917** First recoilless gun, the Davis cannon, is used from aircraft against submarines.
- **1925** The Oerlikon aircraft cannon.
- **1931** Germany begins development of rockets for air defence and long-range bombardment.
- **1932** First studies in biological warfare—the use of disease germs—in Europe, the USA, and Japan.
- **1936** US Army adopt the Garand automatic rifle, the first such weapon to become standard infantry issue.
- **1936** Discovery of nerve gases in Germany.
- **1938** British defensive radar system goes into operation.
- **1940** First use of the shaped-charge principle, in demolition charges against Eben Emael fort in Belgium.
- **1940** The British use a shaped-charge rifle grenade.
- **1940** The British introduce air defence rockets.
- **1940** The Germans introduce lightweight recoilless guns for field use.
- **1942** The Germans develop the "assault rifle".
- **1942** US development of the shoulder-fired "Bazooka" anti-tank rocket.
- **1942** Germany introduces rocket-boosted artillery shells.
- **1943** First use of proximity fuse against aircraft by US Navy.
- **1943** First guided aircraft bombs.
- **1944** Germany introduces pilotless flying bombs ("V1").
- **1944** Germany introduces ballistic missile (the "V2" rocket).
- **1944** First air-defence guided missiles are developed in Germany.
- **1944** First wire-guided anti-tank missile is developed in Germany.
- **1945** First test explosion and military use of atom bomb by the USA against Japan.
- **1950** First military use of helicopters, for casualty evacuation in Korea.
- **1954–73** Vietnam War, use of chemical warfare (defoliants and other substances) by the USA.
- **1955** First strategic guided missiles (Soviet SS-3) enter service.
- **1965** First anti-missile missile (the US "Sprint").
- **1982** Anti-ship missiles are first used in the Falklands War.
- **1983** Star Wars or Strategic Defense Initiative research is announced by the USA to develop space laser and particle-beam weapons as a possible future weapons system in space.
- **1991** "Smart" weapons are used by the USA and allied powers in the Gulf War; equipped with computers (using techniques such as digitized terrain maps) and laser guidance, they reach their targets with precision accuracy (for example a "smart" bomb destroys the Ministry of Air Defence in Baghdad by flying into an air shaft).